# Methods in Futures Studies

## A Westview Replica Edition

The concept of Westview Replica Editions is a response to the continuing crisis in academic and informational publishing. Library budgets for books have been severely curtailed. Ever larger portions of general library budgets are being diverted from the purchase of books and used for data banks, computers, micromedia, and other methods of information retrieval. Interlibrary loan structures further reduce the edition sizes required to satisfy the needs of the scholarly community. Economic pressures (particularly inflation and high interest rates) on the university presses and the few private scholarly publishing companies have severely limited the capacity of the industry to properly serve the academic and research communities. As a result, many manuscripts dealing with important subjects, often representing the highest level of scholarship, are no longer economically viable publishing projects--or, if accepted for publication, are typically subject to lead times ranging from one to three years.

Westview Replica Editions are our practical solution to the problem. We accept a manuscript in camera-ready form, typed according to our specifications, and move it immediately into the production process. As always, the selection criteria include the importance of the subject, the work's contribution to scholarship, and its insight, originality of thought, and excellence of exposition. The responsiblity for editing and proofreading lies with the author or sponsoring institution. We prepare chapter headings and display pages, file for copyright, and obtain Library of Congress Cataloging in Publication Data. A detailed manual contains simple instructions for preparing the final typescript, and our editorial staff is always available to answer questions.

The end result is a book printed on acid-free paper and bound in sturdy library-quality soft covers. We manufacture these books ourselves using equipment that does not require a lengthy make-ready process and that allows us to publish first editions of 300 to 600 copies and to reprint even smaller quantities as needed. Thus, we can produce Replica Editions quickly and can keep even very specialized books in print as long as there is a demand for them.

# About the Book and Authors

## *Methods in Futures Studies: Problems and Applications*

Brita Schwarz, Uno Svedin, and Björn Wittrock

The frequent appearance during the past decade of a variety of futures studies has raised a number of philosophical and methodological issues, as well as criticism of specific approaches. Is it in fact possible to study the future at all? How useful or scientific are the methods commonly used? Addressing these and similar questions, the authors of this book present three examples of futures research conducted at the national level of policy and decision making in Sweden. The examples illustrate the problems of applying knowledge during the course of a futures research project, the ways in which different methodologies interact, and various means of combining and adapting methodological tools and techniques. On the basis of comparisons of the studies, the authors discuss more general issues, including the role of prediction in futures studies, the discipline's impact on policy making, the usefulness of formal models and scenarios, and the problems of establishing research criteria.

Although the authors emphasize that it is impossible to specify, much less prescribe, unique solutions to many methodological problems, they suggest guidelines for the conduct of futures studies, stressing that the value of any method depends on a particular project's organizational and policy-making framework and its objectives. The book includes a concise overview of the characteristics and dimensions of futures studies and the methods and analytical techniques available.

Brita Schwarz is director of research in systems analysis and planning at the Research Institute for National Defense and is also associated with the Economic Research Institute at the Stockholm School of Economics in Sweden. Uno Svedin is assistant professor of physics at the University of Stockholm. Björn Wittrock is associate professor of political science also at the University of Stockholm.

# Methods in Futures Studies
## Problems and Applications

Brita Schwarz, Uno Svedin,
and Björn Wittrock

Westview Press / Boulder, Colorado

A Westview Replica Edition



Published in 1982 in the United States of America by
Westview Press, Inc.
5500 Central Avenue
Boulder, Colorado 80301
Frederick A. Praeger, President and Publisher

Library of Congress Cataloging in Publication Data
Schwarz, Brita, 1933-
Methods in futures studies.
(A Westview replica edition)
1. Forecasting--Methodology. I. Svedin, Uno, 1943- . II. Wittrock, Björn. III. Title.
CB158.S288 003'.2 82-2825
ISBN 0-86531-909-X AACR2

Printed and bound in the United States of America

# Contents

# Preface

Futures studies have been appearing with increasing frequency during the past decade. Some of them have included dire warnings about the doom of mankind; others have painted pictures in the brightest of colours, showing futures of unbounded growth. Some have spotlighted the future of the world for a century or more, others have prefigured the road ahead for single nations or smaller local communities for perhaps the next ten or twenty years. Some have triggered off widespread public discussion and aroused considerable dissent; others have seen their results fed into opaque planning systems with little public ado.

The reactions to these developments, and the methodological debate in general, have touched upon several fundamental philosophical issues and also included criticisms of specific approaches. Is it in fact possible to "study" the future at all? What was "scientific" about the methods used by the authors of "Limits to Growth" that so greatly impressed the general public? Has not the usefulness of the scenario method been exaggerated?

The authors of the present volume, all of us representing quite different disciplinary and professional fields, found ourselves discussing problems of this kind together. It soon became clear to us in the course of our internal talks that often it is impossible to specify, let alone prescribe, unique "solutions" to the methodological problems of futures studies. Nevertheless, the basic thrust of our argument is that some guidelines for the conduct of such studies can be established, but the usefulness and viability of any methods will depend crucially on the organizational and policy-making framework in which any particular project is embedded as well as on the objectives that the futures study is intended to serve.

We thus try first to summarize some of the defining characteristics and dimensions of futures studies. We then illustrate in considerable detail the way in which different methods have interacted in some real-world projects in a specific organizational and policy-making environment, namely at the national level of policy and decision-making in Sweden. We have tried to concentrate on problems of some general relevance in a range of different settings. A central focus in our inquiry has been an examination of three specific projects with which we have been separately involved over a number of years or have even participated in for the whole of the particular project's duration.

Although our original purpose was to tackle some of the general issues arising in the debate about futures studies, we believe that the resulting product may in fact be useful in several areas and in several ways. Since we have tried to launch the discussion from a clear conceptual base, the present book could find a useful place in university courses in futures research. The case histories of the three futures studies and our evaluations of these could be of interest to the organizers and users of futures studies and possibly also on a more general level to anyone interested in interdisciplinary research. We hope, too, that the same applies to the more general analysis of adequacy criteria and the relationship between expert knowledge and public policy.

As the present text has evolved from the gradual merging of individual contributions, it is not possible to identify exactly the origin of the various parts. Each author, however, was responsible for the first rough outline of one of the cases in Chapter 4, which may explain some of the remaining differences in style. Chapter 5, on the other hand, was very much the result of a lengthy process during which ideas and contributions from all three authors converged and merged.

For comments on a previous draft we are indebted to participants in a seminar on "Methods of Futures Studies". We also owe a debt of gratitude to Nancy Adler who has helped to unravel some of the worst linguistic tangles in the final version of the manuscript.

# 1
# Introduction

There is a growing literature on the methodological issues of futures research. In recent years many examples of futures studies have also been published, ranging from global models of the world's future to studies of more limited problem areas. There is, however, a certain gap between these two kinds of literature, as the first usually deals separately with each type of method and the second seldom discusses the choice of methods in any detail. Ways of combining the various methods and approaches and the relationship between the choice of method and the organizational conditions are problem areas that as yet have recieved little attention. Nevertheless, such aspects can be important both to people working in the field of futures studies and to the users of their studies. This book represents an effort to bridge the gap. More specifically, it is the result of a discussion between the authors of their own experiences and studies in this area.

As a general introduction we briefly describe the characteristics of futures studies in Chapter 2. The type of problem area, the purpose and goals, and the available knowledge and methodological tools all combine to determine the boundaries of such studies. Futures studies represent a new field of intellectual endeavour and the development of a methodology in the field has not been in any way exceptional. It is thus not the methods in themselves that seem to us to merit special attention, but rather their focus: they are interesting because of the problems with which they are concerned, namely the future and the possibility we have of influencing it.

In Chapter 3 we give a brief exposé of some of the methods and analytical techniques available. Terms such as scenarios, mathematical models etc have rather varying connotations in the methodological debate, and we therefore try to clarify our own use of these terms.

The advantages and disadvantages of specific techniques such as mathematical modelling and Delphi are frequently debated. We have chosen to examine questions of this kind on a basis of actual cases. We use the term methods of futures studies in a broad sense and discuss problems that arise in applying and organizing knowledge and competence during the course of a project, as well as various ways of combining and adapting tools and techniques. In Chapter 4 three case studies are presented. They describe how these methodological issues have been treated in three futures studies published in Sweden during the past two or three years. The studies concern energy, global resources and transportation systems. A reason for selecting these three futures studies was that they display enough similarities and dissimilarities to permit a fruitful comparative study.

In Chapter 5 the three case studies are compared. The discussion is geared to certain general issues and among other things we argue that:

- Prediction is not the main role of futures studies.
- The organizational setting of a futures study will affect the general conduct of the study as well as the choice of methods and techniques.
- In a context of planning and policy-making, futures studies cannot be regarded as mere inputs into neatly ordered planning and decision-making systems. Often their most profound impacts are indirect, and sometimes conditioned by political controversy.
- A fruitful way of approaching methodological choices may be by considering the acquisition and synthesis of various kinds of knowledge. Problems connected with the acquisition and integration of knowledge may well necessitate awkward choices between the risk of succumbing to methodological rigidity and the risk of failing to present a coherent analysis.

- The discussion of the alleged superiority of formal models or of scenarios rests on misconceptions.
- Finally, we agree with the commonly held view of the crucial role of values in futures studies, but argue that the focus has been too much on overtly political values at the expense of less salient but potentially more important values of a professional and paradigmatic nature.

In a concluding chapter we bring together some of the main themes of the entire volume in a discussion of three aspects of the legitimacy of futures studies: a section on <u>intellectual legitimacy</u> focusses on criteria of quality and adequacy, another focusses on <u>professional legitimacy</u> and summarizes some key professional roles that futurists might occupy, finally, we return to the <u>societal</u> and <u>political legitimacy</u> of futures studies as one of the many forms of social problem-solving.

# 2
# Characteristics and Problems of Futures Studies

Interest in the future is nothing new. Prophecy seems to be an integral part of many religious, literary and political practices. However, the emergence of futures studies as a special field of interest for research institutions, governmental commissions and professional journals is largely a phenomenon of the last two decades. Futures studies can be characterized by the general ambition to support decision-making by supplying information of relevance in a long-range perspective (Boucher, 1977, pp.6-9). Thus, futures studies cannot be associated exclusively with any specific technique of analysis; they should be regarded rather as a task encompassing a series of activities and often calling for a broad approach. Nevertheless, these studies exhibit a number of distinguishing features and give rise to certain special problems. Generally speaking the function of futures studies, namely to provide information relevant in a very long-range perspective, accentuates many problems already familiar from various forms of policy analysis and planning. Important characteristics of futures studies can be described in the following dimensions.

(1) Time horizon. Sometimes the time horizon of futures studies is indicated in terms of a time span, perhaps "from five to fifty years" (or sometimes even longer), or in terms of a time period such as "beyond the next two or three decades". However, it is not the time horizon as such which is the distinguishing feature of futures studies; it is rather the emphasis on the way conditions and situations change over time. The chosen time horizon is often so distant that a number of different "futures" are possible, and the alternative that emerges will be dependent on decisions made today. In some Swedish futures studies a time horizon defined by "the relative lack of major commitments so that consideration can be given to qualitative changes" has been suggested, thus implying an obvious

orientation towards policy-making and action.

(2) The problems which the futurists treat tend to have an open-ended character in at least three respects. (i) The problems do not often fall directly within the area of responsibility of any existing administrative or policy-making unit, but instead cut across several boundaries - the more fuzzily the more distant the time horizon. (ii) An interdisciplinary approach is often recommended in connection with complex real-world problems. For emerging problems of the future there is an even more pronounced need to draw upon competence from a number of different disciplines, and to utilize any kind of knowledge which might bear on the problem. (iii) The problem formulation itself may constitute a major part of the analysis, and definite solutions cannot usually be expected. In other words futures problems are wicked problems, to use the Rittel-Webber distinction between tame and wicked problems (Rittel & Webber, 1973). Thus, in several important fields of interest to futures studies, well-established scientific theories are simply non-existent. Our knowledge is partial and will have to remain so, although some improvements are possible.

(3) Uncertainties and discontinuities. Uncertainties are common both in everyday life and in many problems and processes that are empirically studied. However, we generally know what variables we are interested in, i.e. the dimension, and we may also know something of the probability distribution, if the phenomenon is of a statistical nature. Usually the term risk is used when we know the probability distribution but not the outcome. A situation in which the dimension is known but not the probability distribution is sometimes said to exhibit pure or "secondary" uncertainty (Dror, 1971, p. 64; Strangert, 1977, p. 153). In a long-term perspective several of the assumptions underlying short- and medium-term forecasting may become invalid, and new variables often become important. We are thus sometimes faced

with "primary" uncertainty, i.e. when the dimensions of possible results are unknown. In the case of pure or primary uncertainty we are faced with the problem of proceeding "with the identification of discontinuous changes and innovations" (Shani, 1974, p. 643). Some authors claim this kind of identification to be a main element of futures studies. In this context it may be illuminating to recall that one of the origins of what is nowadays termed futures studies was the development of the Delphi technique, and its use for obtaining the opinion of a group of experts about possible future technological breakthroughs and their probable occurence in time.

(4) Criterion of adequacy. The future is uncertain because of our limited knowledge of the world in general and because the future itself will evolve as a result of decisions not yet made. Futures studies do not therefore aim primarily at producing predictions of future events; their aim is rather to provide an overall picture which is relevant to the problem at hand. Hence futures studies often deal with the design of alternative futures. The picture of the future given in a futures study has to be adequate in the sense that it permits the identification of "thresholds which would dictate that qualitatively different allocation of resources should be made" (Cole, 1976, p. 310).

(5) The role of theory. It has already been noted that futures studies are often unable to rely on established theories defining relevant sets of causal relations. This, however, should not be taken to imply that futures studies are inherently atheoretical in the sense that they use some sort of "naive" forecasting methodology (Edman, 1976). But the hypothetical nature of many of the relationships and sequences dealt with in futures studies seems to make futurists feel more free in the choice of theoretical assumptions than traditionally appears to be the case among most social scientists. Futures studies thus involve an effort not only to use but also to reconstruct theories and theoretical

assumptions. In this way futures studies may have an effect on the development of theories within various disciplines.

(6) Techniques of analysis. It is sometimes suggested that futures studies involve the use of so-called intuitive forecasting techniques in contradistinction to more formal and rigorous methods. Examples of such intuitive techniques are scenario construction, Delphi techniques, brainstorming and morphology. However, the utilization of intuitive techniques does not necessarily exclude the use of formal or quantitative techniques within the same study. On the contrary, some of the more interesting experiences, at least in Swedish futures studies, have been connected with a combination of, for example, normal econometric modelling and scenario writing. Moreover, various general approaches, such as systems analysis and policy analysis, which are often associated with futures studies, incorporate the use of analytical techniques of both the intuitive and the more rigorous kind.

(7) Proximity to the application of results. The proximity to the application of the results of futures studies may vary considerably. In Sweden the Alva Myrdal commission on futures studies made a distinction between three categories which can be defined as follows: (1) studies that may serve fairly directly as inputs into existing public planning and decision-making; (2) research which may indirectly help to improve future planning and decision-making processes; and (3) autonomous futures studies governed neither directly nor indirectly by the requirements of existing government bodies.

(8) Linkage to planning processes. Hence, futures studies may be geared more or less closely to planning and decision-making processes. Planning is sometimes described as a process involving the preparation of a set of decisions such that a "system" or "planning object" can be adapted to its "environment"

in a way that is consistent with prevalent goals and values (see e.g. Ingelstam, 1977, pp. 127-145). From this description it thus follows that planning has to be based on information about the future development of the environment of the planning object. Studies of such developments can be considered as a form of futures studies. In this case the role of futures studies is to generate direct inputs into the planning process.

However, long-range planning sometimes involves the design of alternative futures for the planning object itself. When this is the case, the actual planning can be regarded as a form of futures study, so that there is no clear borderline between the two. However, Shani has developed a conceptual framework intended to distinguish between planning and futures studies and which thus seems relevant to an exploration of their relationship (Shani, 1974). As far as organizational matters are concerned, Shani argues that futures studies "should be liberated from ways of thinking conditioned by existing, organizational patterns. Thus, while in planning reference to the future would have inevitably to be from the organization outwards, in futures studies the orientation would be from the environment to the organization". Needless to say, these orientations refer primarily to shifts in emphasis: any realistic process of planning or futures studies will have to consider the interplay between environment and organization from both perspectives.

(9) Values, goals, and norms. It is often said that values and norms should be explicitly stated in futures studies - a recommendation that is partly related to the idea that futures studies should, directly or indirectly, influence planning and policy-making. However, given the complexity of planning processes and the changeability of values and goals over time, the role of values in futures studies cannot but be ambiguous. It is sometimes argued that the generation of alternative futures from widely different value-premises may lead to insights into the

range of real uncertainty (Cole, 1976, p. 314; Clark & Cole 1975, p. 114). Experiences from the Swedish studies described in Chapter 4 indicate, however, that the nature of the more precise relationships between value dimensions and the actual activities of futures studies is far from clear. This remains true even if we restrict ourselves for instance to some specific version of scenario-writing. Hence, different values and norms may be considered in each scenario or all scenarios presented may involve the satisfaction of certain values. Furthermore, in some scenarios the values and norms of a group or of a whole society may be assumed to change and the assumed implications of such changes can then be described in the scenario. Finally, and perhaps of most importance, is the role that values may play as a source of discussion and conflict within a given project group. And this may be so even if the group never actually engages in any kind of explicitly normative futures studies (cf. Wittrock, 1979).

## REFERENCES


BOUCHER, W.I., "Introduction", in W.I. Boucher (ed), The Study of the Future: An Agenda for Research (Washington, D.C., National Science Foundation, 1977).

CLARK, J. & S. COLE, Global Simulation Models - A Comparative Study (John Wiley & Sons, London, 1975).

COLE, S., "Long-term Forecasting Methods: Emphasis and Institutions", *Futures* 8 (August 1976).

DROR, Y., Design for Policy Sciences (Elsevier, New York, 1971).

EDMAN, M., "There is no naive way of making forecasts", in S. Schwarz (ed), Knowledge and Concepts in Futures Studies (Westview Press, Boulder, Colorado, 1976).

INGELSTAM, L., "Basic problems in planning" in Trends in Planning, edited by C-G Jennergren, S. Schwarz, O. Alvfeldt (Wiley, London, 1977).

RITTEL, H.W.J. & M.W. WEBBER, "Dilemmas in a General Theory of Planning", *Policy Sciences* 4 (1973), pp. 155-169.

SHANI, M., "Futures Studies Versus Planning", *OMEGA* 2 (1974:5) pp. 635-645.

STRANGERT, P., "Adaptive Planning and the Resolution of Uncertainty" in C-G Jennergren, S. Schwarz, O. Alvfeldt (eds) *op.cit.*, pp. 147-165.

WITTROCK, B., "Long-range Forecasting and Policy-Making" pp. 267-288 in T. Whiston (ed) The Uses and Abuses of Forecasting (Macmillan, London; Holmes and Meier, New York, 1979).

# 3
# Methods of Futures Studies

The emergence of futures studies as a distinct field of inquiry in the 1960s and 1970s was interwoven with the development and increasing sophistication of various analytical techniques such as Delphi, scenario construction, dynamic mathematical models, and various forms of cross-impact analysis. Some of these techniques appear to be of a highly formalized kind, whereas others seem to depend heavily on intuitive judgements. In the following chapters we will argue that these various techniques should not necessarily be contrasted and regarded as mutually exclusive. Actually, from both a methodological and historical point of view, it is rewarding to focus on the interplay between so-called intuitive and formalized techniques. Thus, we will limit ourselves here to a brief review of some types of techniques, starting with Delphi and the well-known criticism to which it was subjected by Sackman. We proceed with trend extrapolation and discuss what we consider to be misunderstandings about its use in forecasting. We then examine in more detail some problems connected with the scenario method and mathematical modelling. Apart from permitting us to elaborate on the combination of apparently quite distinct and mutually exclusive techniques, these last two have been utilized within the framework of the futures studies projects presented in Chapter 4.

Method literally means "proceeding towards" something. We will use the term as a generic concept encompassing not only techniques of analysis but also problems pertaining to the utilization and organization of knowledge and competence during the course of the study and in the presentation of the final results. Such a delimitation is justified on several counts; it is often impossible to treat futures problems unless a trans- or multidisciplinary approach is used, and unless competence from different fields is organized effectively. Linking futures studies to planning and decision-making processes is not only or even primarily a matter of providing results to serve as inputs into the processes; it is a question of organizing futures studies in such a way that experiences of new models of thought can be transferred to planners, for example, in the course of a futures studies project. Some of the specific techniques of futures studies, e.g. Delphi techniques and to a certain extent cross-impact analysis, are deliberately designed to provide for an exchange of views from different fields of knowledge, i.e. they function as methods for organizing and utilizing competence.

## 3.1 THE DELPHI TECHNIQUE

Many versions and variants of Delphi have been developed since the technique was first designed and applied at the Rand Corporation (California, USA) in the early 1950s. It originated as a method for eliciting expert opinion by means of a structured communication process. It became more widely known after the publication in 1964 of an application to long-range technology forecasting (Gordon & Helmer, 1964).

Typically, conventional Delphi involves a monitor team that explores the problem area and designs a questionnaire. This is then sent to a panel, a group of experts selected by the team. The monitor team summarizes the results of the responses and designs a new questionnaire for the group of experts. The experts are usually given at least one opportunity to re-evaluate their original answers, after having received information about the group response. This is often given as an average value together with some measure of dispersion. Emphasis is put on anonymity, i.e. the individual responses of the experts are not communicated to the others, nor are they published. Earlier forms of Delphi often aimed at a consensus among the experts.

The most common form of Delphi is probably the paper-and-pencil version described above. However, there is a more recent form, sometimes called a "Delhpi Conference", which makes use of a computer programmed to compile the group results. The emphasis here is on the communication process. While Delphi was originally used for forecasting purposes, it has spread to many other areas of inquiry during the last decade. It has been used, for instance, in exploring the advantages and disadvantages of available policy options and in exposing priorities among personal values and social goals. Depending on the issues explored, the panelists may be experts or laymen. First used by futurists, Delphi has now spread to new user categories such as social scientists in general; it is also used quite often by administrators. Delphi no longer necessarily aims at consensus; it has also been used to explore differences of opinion between groups of people (see e.g. Wikström, 1974).

Sackman has criticized the Delphi technique for being less rigorous than other questionnaire techniques (Sackman, 1974). However, futurists have claimed that it is "a method of last resort", a method which can be considered when no other methods are available or when other methods would be too expensive (Coates, 1975). In a survey of technology forecasting methods Martino starts with Delphi, presenting it as the most intuitive of the technology forecasting methods (Martino, 1973). And it has also been regarded as mainly a brainstorming device. To quote from Sackman (1974, p. 71):

> "Nanus, Wooten, and Borko (1973) in a relatively ambitious Delphi study on the social impact of the multinational computer, make it clear that no claims are made for the reliability or validity of their Delphi results - Delphi was used for strictly exploratory purposes in an uncharted domain, "... the authors chose to use Delphi with realization that the results would be more in the nature of a structured 'brainstorming' session with noted thinkers than a scientific exercise in prediction" (p. 11).

In a recent survey of the Delphi method (Linstone & Turoff, 1975, p. 4) the variety of application areas is mentioned, but it is stressed that it is not

> the explicit nature of the application which determines the appropriateness of utilizing Delphi, rather, it is the particular circumstances surrounding the necessarily associated group communication process. ... Usually, one or more of the following properties of the application leads to the need for employing Delphi:
>
> . The problem does not lend itself to precise analytical techniques but can benefit from subjective judgements on a collective basis
>
> . The individuals needed to contribute to the examination of a broad or complex problem have no history of adequate communication and may represent diverse backgrounds with respect to experience and expertise
>
> . More individuals are needed than can effectively interact in a face-to-face exchange
>
> . Time and cost make frequent group meetings infeasible
>
> . The efficiency of face-to-face meetings can be increased by a supplemental group communication process
>
> . Disagreements among individuals are so severe or politically unpalatable that the communication process must be referred and/or anonymity assured
>
> . The heterogeneity of the participants must be preserved to assure validity of the results, i.e. avoidance of domination by quantity or by strength of personality ("bandwagon effect").

Although Delphi played an important role in the evolution of futures studies, it is by no means the dominant tool of the futurists. However, in view of the focus of this study, it is interesting to note that it has sometimes been used in combination with other methods (see e.g. Wildhorn, 1976).

In none of the three case studies outlined in Chapter 4 was the Delphi method used. However, an interesting aspect of Delphi is that it addresses a central issue in futures studies, namely the necessity of eliciting often quite vague forms of knowledge from a variety of people.

## 3.2 TREND EXTRAPOLATION

To regard trend extrapolation as a method of futures studies is perhaps somewhat paradoxical, since the start of the "futures studies movement" in the 1960s can be seen as a reaction against trend extrapolations, or perhaps against the view that the future is uniquely determined by past trends. This has been stressed, for instance, in arguments about the use of terms such as futures studies, futures research and "futuribles" rather than futurology or future research. Bertrand de Jouvenel has suggested the term "futuribles" to distinguish a future state made plausible or imaginable by new developments (Jouvenel, 1967). Calder has argued that the term "futures research" is better than "future research" because "it conveys a crucial point: the aim is not to prophesy what the actual future will be but rather to rehearse futures accessible to political choice" (Calder, 1967).

Nevertheless, trend extrapolations are frequently used in futures studies, but usually only in connection with certain quantifiable aspects or parts of the problem. When we talk about futures studies we usually mean the study of a very large problem area or a complex social system; examples would be studies of the future competitiveness of Swedish industry or studies of the future development of the energy sector or transportation sector in Sweden. Trend extrapolations may be used here for certain variables such as population growth in urban areas, productivity increases in the transportation of goods, energy consumption etc, but probably only at an introductory stage to obtain some reference projections.

To discuss the role of trend extrapolations we need to examine the kind of knowledge that is required to make an extrapolation and, equally importantly, what kind of "knowledge" can be gained from such an extrapolation. Obviously some kind of historical data is needed. Assume that we have a time series of data as below.

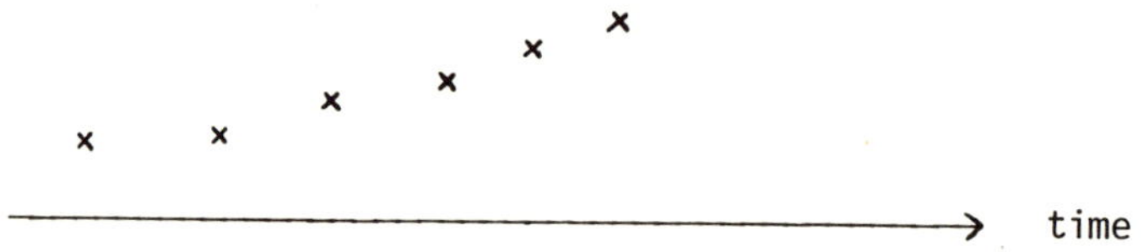


The simplest kind of trend extrapolation involves plotting the data as a function of time, the fitting of a line or curve identifying the trend, and then an extrapolation of the trend as in the figure below.

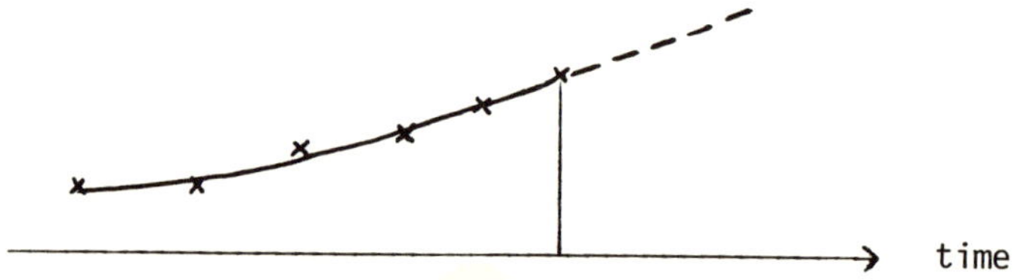


But when can such an extrapolation be considered an accurate forecast, i.e. a well-founded proposition about the future? We have encountered the following "ideas" related to this issue, all of which we think reveal a misunderstanding of trend extrapolation.

A. The "no-understanding" definition

According to this idea, trend extrapolation is used when we make a forecast of the future development of a variable by extrapolating a trend that can be identified in one way or another from time series data, when we know nothing about the causes of the change and possess no supporting knowledge about the variable other than the historical data.

B. Good fit

According to this idea, the possibility of fitting historical data to a mathematical function f (t) is of overriding importance and is a sufficient condition for regarding f (t) as a reliable scientific prediction of the future development of the variable. The emphasis is thus on the situation in which the historical data displays a considerable measure of regularity.

C. Earlier successes

If repeated forecasts of future outcomes of a certain variable have been made, using the same trend extrapolation procedure, and if all the forecasts have turned out to be correct, then - according to this idea - we have "validated" the procedure. In this case the trend extrapolation method is sometimes believed to provide valid scientific predictions.

We will now comment in turn on each of these three "misunderstandings".

### The "no-understanding" definition

First of all there is nothing clear-cut about the actual making of the fit; what curve is to be used? Even to be able to answer this question we have to have some idea of the process involved. Is it essentially a smooth function or does it fluctuate, is it cyclic or non-cyclic, etc? If we know nothing about the phenomenon except for some historical data, then we really have no reason to believe that a trend extrapolation will be useful as a forecast. Assume for instance that a man from another planet visits the earth for the first time in his life and that he starts counting the numbers of cars passing a road early one morning between 6 and 8 o'clock getting the following result:

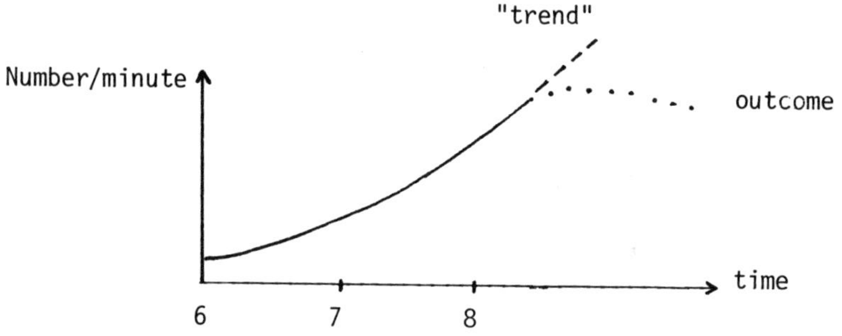


As he knows nothing about the causes of traffic or of normal traffic variations during the day, it would obviously be unwise of him to try to make a forecast by using trend extrapolation. In this case the direct cause of error is his ignorance of the fact that traffic normally levels off towards 9 o'clock. Thus, people using trend extrapolation for forecasting purposes usually have more information about the penomenon than just a data time series, but their understanding of the causes of change may be incomplete, nebulous or even subconscious.

## Good fit

When historical data exhibits a marked regularity (see example below), it is usually easy to find a mathematical function f (t) that fits. Such a function then represents a good description of the available data. But if we have no knowledge whatsoever about the underlying causes, we cannot trust f (t) to provide an accurate forecast.

Take for instance world consumption of coal. For a number of years there has been a steady relative decrease in coal consumption. See figure below. But to extrapolate this trend over a

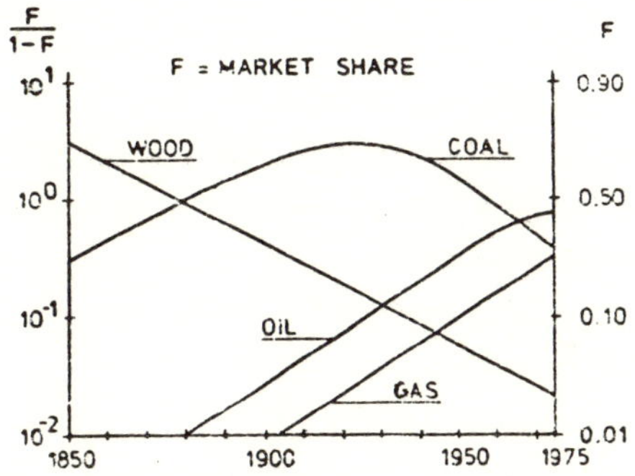


Fig. 3:1. Global market shares (F) of primary energy forms.

lengthy period of time would be unlikely to result in a good forecast. Increasing oil prices and various problems connected with other energy sources are likely to affect and probably reverse the earlier downward trend in coal consumption.

## Earlier successes

Even when trend extrapolation has been used for forecasting several times with success, this cannot be regarded as a "validation" of the method. For the same reason that a good fit with historical data is no guarantee that trend extrapolation will be useful in forecasting, earlier successes are no guarantee against failure. If the underlying causes of change are unknown, or if we are mistaken about them, or - and this is not unusual - - if new causes of change enter the picture, a trend extrapolation may result in a forecast that is quite misleading.

## Discussion

These three cases illustrate the importance of a critical attitude towards trend extrapolation. But the problem remains. What information is needed if we are to use trend extrapolation for forecasting? A rather obvious requirement is that the forecaster should have some idea of the factors that have caused changes in the past; he should also have good grounds for believing that they will continue to influence developments similarly in the future. How much he needs to know, and in what detail, depends on the time span intended for the forecast and the degree of accuracy required.

We have only mentioned the most simple trend extrapolation technique; to plot a curve, to identify a trend visually, and to extrapolate from it. There are also a number of more sophisticated techniques. Most of these require some knowledge of the possible causes of past fluctuations and changes. For instance, are there seasonal variations? Or are there certain mere chance fluctuations, which should be removed from the data before a trend is identified? Methods for describing and projecting a trend include linear or curvilinear regression analysis, and the Box-Jenkins methods for time-series analysis (Box & Jenkins, 1970). Including explanatory variables may or may not lead to improved forecasts, see Gärdenfors (1980).

In any futures studies in which trend extrapolation is being used or considered, the phenomenon to be studied is usually very complex (coal consumption, yearly passenger kilometres by car etc). It is also subject to the influence of a great many factors, including decision-making, about which our knowledge is necessarily very incomplete. Consequently the purpose of the trend extrapolation in this type of case is not usually to get "accurate predictions" of future outcomes but rather to get a reference projection which can serve as a starting point for further investigations. When a trend is estimated from time series data and projected into the future, the projection theoretically answers the following question: if this data series continues into the future for a certain number of years in accordance with the pattern assumed in the trend extrapolation, what values will the variable attain at different points in time during this period?

When can a trend extrapolation be expected to yield a "rational scientific forecast"? M. Edman has examined this issue in a philosophy of science context in his paper "There is no naive way of making forecasts" (Edman, 1976). Edman argues that when several possible generalizations are supported by existing evidence, it is the more "lawlike" and not necessarily the simplest generalization which should be regarded as the rational conclusion.

Obviously the kind of variables which are important in futures studies - such as coal consumption, urban population, etc - may develop differently in the future than they have done in the past because of new events, new causes of change, new policies etc. A criterion of "law-likeness" may thus be able to help to identify an existing trend, but there is no guarantee that such a trend will continue. In forecasting it is very important to consider new causes of change and this sometimes creates a responsibility problem. If the "new factor" is outside the field of expertise of the forecaster, he may not feel obliged to consider it. But the user of the forecast may be expecting the forecaster to have taken "everything" into account. An example of this type of problem is discussed by Nanus in "Interdisciplinary Policy Analysis in Economic Forecasting", in which an economist is quoted as having argued that economists cannot be held responsible for not having foreseen the part of the double-digit inflation of 1974-1975 which depended on political and meteorological factors (Nanus, 1979, pp 288-289).

In an overview of trend extrapolation in futures studies, Hill (1976) has pointed out the usefulness of the method as a "baseline" technique appropriate to an introductory stage. He says:

> "Because of its intuitive appeal and low cost, a trend study can be rather quickly completed and interpreted. If interesting new problems or hypotheses are brought to light by this method, the analyst can address such issues subsequently with more rigorous - and more costly - approaches. When trend methods are employed in this manner, they will usually result in such speculative questions for further research as
>
> - What will be the consequences if this trend continues unabated for x years?
> - What will be the consequences if this trend levels off or accelerates from its current rate of increase?
> - Which forces acting to maintain this trend are likely to remain constant and which are likely to change in the future?
> - If we wish to alter this trend in a particular direction, what determining variables might themselves be manipulated to do so?"

Trend extrapolation is often used in futures studies in combination with other methods. It may be used as an element in a scenario. More complex trend extrapolations may involve the development and use of a mathematical model.

## 3.3 SCENARIO-WRITING - HISTORICAL BACKGROUND

Scenarios are regarded by many futurists as one of the main tools of futures studies. However, the increasing use of scenarios of various types and for different purposes has made the term scenario become almost synonymous with the whole area of futures studies. This naturally poses something of a semantic problem for anyone trying to assess the use of scenarios as a methodological approach. The variety of connotations attaching to scenario can probably best be understood in light of the historical development of the term, beginning with the original meaning of scenario and its first connection with planning and futures studies. Below we examine this development and identify a number of methodological ideas which may be in people's minds when they refer to the scenario method.

According to the Oxford dictionary a scenario is "a sketch or outline of the plot of a play, giving particulars of the scenes, situations etc". From this definition it seems rather far-fetched to use the word scenario to mean a description of a future development. However, there is a connection. The explanation seems to be the increasing use of gaming in operations research and planning during the 1950s. Particularly in defence, gaming was used as a tool in long-range planning. New weapon systems were "tested" in map games before the final decisions were taken about development and acquistion. These games were therefore simulations of hypothetical conflicts 10 to 20 years ahead and the scenarios were "the outline of the plot" made before a gaming exercise was started. Such scenarios usually consisted of a verbal description of a hypothetical situation at a future point in time; but they could include a sketch of the main changes which were assumed to have taken place in the intervening period.

Herman Kahn, who used scenarios in the 1950s in connection with defence studies at Rand in the U.S, later made the term familiar to a great many people through his well-known book "The Year 2000" (Kahn & Wiener, 1967). There he defines a scenario as "a hypothetical sequence of events constructed for the purpose of focusing attention on causal processes and decision points". He presents a number of different "scenarios", consisting of macro-level descriptions of alternative developments of society towards the year 2000. An interesting methodological development here was the design of a so-called "surprise-free" scenario, a development in which present trends continue. Such a scenario can be useful as a reference projection, a starting point for the construction of alternative scenarios. As Kahn points out, a "surprise-free" development is not very likely; history has produced few periods of any length without various breakthroughs or turning-points. Nevertheless the design of a "surprise-free" scenario as a reference has often proved to be a convenient step in scenario work.

In 1968, in a survey entitled "Scenarios in Systems Analysis", Brown states that "there are many notions floating around of what a scenario is or ought to be" (Brown, 1968). Some of the notions mentioned were:

- an outline of a sequence of hypothetical events
- a record of the actions and counteractions taken by parties to a conflict
- a specific set of parametric values selected for a given run of the computer.

Brown notes that "they all - disparate as they are - refer to descriptions of the conditions under which the systems they are analysing, designing or operating are assumed to be performing". Drawing his examples mainly from the defence field, Brown explains that a system may be a weapon, or a vast complex of weapons and support facilities, or an organization; it may even be the entire national security establishment. Brown concludes that whatever the scope and properties of the specific system, a scenario - in systems analysis - can be defined as a statement of assumptions about the operating environment of the particular system which is being analysed.

This discussion of the meaning of the term scenario in systems analysis seems to have a general methodological implication, namely that formal descriptions of assumptions are used or are considered useful in planning. Let us put it another way. When the future development of a system or planning object is studied in planning or systems analysis, it is important to be explicit about the assumptions regarding the future development of the environment of the planning object or "the conditions under which the system is assumed to be performing". In this context the term scenario refers to formalized descriptions of such assumptions. According to Brown such descriptions may or may not be in verbal form, may or may not consist of a set of data, and may or may not include assumptions about how various actors act and react.

At the beginning of the 1970s the study "Limits to Growth" appeared (Meadows et al, 1972). The authors presented a number of global futures based on a systems-dynamics model and obtained with the help of computer calculations. Variables included world population, BNP per capita, environmental disturbances etc. Since these authors did not use the term scenario for the "futures" they presented, mathematical models have to some extent come to be regarded as the opposite or main alternative to scenarios in the methodological debate. Although this view, or should we say this terminology, is not accepted by all futurists, it is discernible in the literature.

In Sweden the use of scenarios became institutionalized in defence at the end of the 1960s when the system for long-range defence planning was reformed. Already at the end of the 1950s, when gaming began to be used as a tool in defence planning, descriptions were made of the assumed environment and the "plot of the play" before the start of a game. Although scenarios, in the original sense of the word, were thus being used, the actual term scenario was not applied to them at that time.

The new system for long-range defence planning was introduced on an experimental basis at the end of the 1960s. One of the elements included in this system consisted of descriptions of a set of alternative future international developments - hypothetical sequences of events - each leading to a crisis which might constitute a threat against Swedish security (Andrén & Schwarz, 1974). Such descriptions are called crisis scenarios. As part of, or in preparation for, the work on the scenarios, special background studies are made by various experts in areas important to the problem area. The work on the scenarios thus acts as a way of communicating knowledge between experts, and of synthesizing it.

A number of acceptability criteria have been developed for the set of crisis scenarios providing a basis for defence planning. Certain requirements as regards consistency, rationality and relevance, for instance, are attached to each scenario.

In addition to the criteria defined for each separate scenario, there are also certain criteria for the set of scenarios. For instance, together the scenarios should define and delimit the variance in the security policy assumptions on which Swedish defence planning should be based. Thus, in a way, the scenarios can be regarded as a set of vectors, together defining a space of future developments which is considered important to the study at hand. In fact, this seems to be a fairly common way of using scenarios, i.e. employing a set of scenarios that together define an area of interest although each scenario separately need not necessarily represent a likely development.

In a recent survey of scenarios, I. Wilson discusses the meaning of the term in futures studies (Wilson, 1978). Among other things he emphasizes the holistic principle. He argues as follows:

> "Scenarios are (or should be) multifaceted and holistic in their approach to the future. In the early days of futures research, and still to a great extent in popular literature, it is the isolated event, the specific prediction that rivets attention. However, history is a 'booming buzzing confusion' of events, trends and discontinuities; it is constantly in motion and so is more accurately represented by a motion picture than by a snapshot. Scenarios have a special ability to represent this multifaceted, interacting flow process, combining (when appropriate) demographic changes, social trends, political events, economic variables, and technological developments."

In a recent review of futures literature and futures forecasting (Rivett, 1979), a scenario is said to be "a qualitative or quantitative description of the world at a particular future time. The scenario describes scene by scene one or more mental images of the future and for the method to be informative it is important that these images be detached, self-consistent and real". The author describes several methods for writing scenarios, one of which is as follows:

> "One writes a description of the world as it might be in the year 2000 and then from the perspective of the year 2000 the history of the previous 25 years is written. In this history, which would take us from the known of 1975 to the suggested world of 2000, one can identify critical points of change in the world which lead to the culmination of the particular history described for the year 2000. By analysing such 'historical' documents one gains an understanding of the events which might have to occur for particular scenarios to obtain in reality. Attention can then be focussed on the likelihood of these events occurring and hence on the likelihood of the particular scenario being evolved."

From the above discussion of the use of the term scenario in connection with futures studies, it should now be clear that the term can have a number of quite different meanings. For instance it may be used to denote a description of:

- a hypothetical, likely or unlikely, development or situation
- a development which is described as caused to some extent by the actions and reactions of various actors
- a desirable or nondesirable development or situation.

And a scenario may also be a verbal description, with or without quantitative information, or a set of parametric values selected for a computer run.

Considering the vague meaning of the term scenario, it may seem surprising that the use of scenarios is regarded as one of the main methods or tools of futures studies. If a scenario is nothing but a verbal description of some hypothetical future development, how can it be regarded as a method? The answer, in our view, is that when people refer to the use of scenarios as a method, they are actually thinking of a number of methodological ideas, some of which we have mentioned here in our sketch of the historical developments. More specifically, we have found in our studies of the use of the scenario method that it usually means applying some of, but never simultaneously all, the following methodological ideas:

a) We cannot predict the future, but by designing a number of plausible and consistent descriptions of hypothetical future developments or situations (scenarios) we can delimit the uncertainty space which we want to take into account in studying the problem at hand.

b) If we want to anticipate problems and improve our future situation, we should not just assume that present trends will continue; instead we should design desirable futures (scenarios), identify branching-points and mechanisms of change so as to find actions which can affect developments in a desirable direction.

c) If we are concerned with the future development of a specific system or planning object (e.g. a business firm, the energy system etc), it is useful to make explicit assumptions (scenarios) about the future development of the environment of the system (otherwise the implicit assumption is often made that there will be no change at all).

d) It may be useful to try to synthesize fragmented, dispersed and sometimes vague knowledge into a holistic and consistent picture (scenario) of a future development or situation. This is particularly important when there are considerable cross-impact effects between developments in various fields which cannot be taken into account if they are considered separately.

e) To get a realistic picture of possible future developments in an area where discontinuities and changing trends may emerge as the result of specific events and actions, it can be useful to try to describe a number of hypothetical developments as resulting in part from the decisions and actions of various actors.

f) Certain future developments are considered unlikely but dangerous. By making an effort to imagine in some detail how such developments might arise, we may be able to make preparations which will render them less probable or less dangerous.

As the term scenario is used nowadays with such a variety of different connotations, it has become less suitable to any discussion of the methodology of futures studies. We can hardly compare the advantages and disadvantages of applying Delphi, scenarios, mathematical models etc in a specific futures study, for instance, when the term scenario may include more or less all the approaches used in this field. It therefore seems necessary to refer explicitly to the set of methodological ideas under consideration, instead of just using the term scenario.

When reference is made to "the scenario method" this may mean the application of some of the methodological ideas listed above. However, it may also involve some specific methodology for developing scenarios. There are a number of such methodologies, the adequacy of which will naturally depend on the type of scenario wanted, the problem area, the organizational setting and so on. (For an example see e.g. Godet, 1979).

## 3.4 MATHEMATICAL MODELS

If scenario writing often appears to represent the "soft" end of the methodology spectrum - together with such approaches as brainstorming and Delphi-techniques - it is at the other end that we find the formal approaches, and in particular the use of mathematical models. While scenarios are often intended to embrace qualitative aspects, mathematical models are concerned with quantifiable features. It has been argued - and in our opinion wrongly - that mathematical models are better than other approaches and that the use of mathematical models represents the most scientific approach.

We do not intend to provide a complete survey or assessment of the role of mathematical models in futures studies. However, given the possibility that unless their role is fully understood the use of mathematical models in futures studies may do more harm than good, it is necessary to examine some aspects of the role of these models. Generally speaking, from a philosophy of science point of view, it needs to be stressed that

- there is no direct analogy between the use of mathematical models in futures studies and their use in the natural sciences
- even within the field of futures studies there are different types of mathematical models and different ways of using them.

First we will comment on some of the differences between natural science models and mathematical models in futures studies, as we think that an underestimation of these differences is one of the reasons why some forecasters claim a higher level of accuracy for the results of their models than the procedures and data allow. We will then present a simple example of a mathematical model which we subsequently use to illustrate various reasons why model outcomes may deviate from the development of the modelled system, e.g. because of new sources of change or human intervention. Finally, we refer to the great variety of mathematical models used in futures studies and emphasize some other purposes besides forecasting for which these models can be used.

## Natural sciences - futures studies

Let us distinguish between some of the different purposes of modelling, beginning with descriptive, explanatory, and forecasting purposes. By fitting a mathematical function to observed data we get a mathematical model which may serve as a practical description of the data concerned. But if the model does not reflect an understanding of the phenomenon observed, it cannot really be expected to serve as a satisfactory tool for forecasting purposes, although it may in some cases be able to generate fairly accurate forecasts, particularly if it is only used for short-term forecasting. If the model is based on some kind of explanation of what has been observed, it is more likely to be useful for forecasting, particularly if it also manages to capture the causal mechanism of change. However, as is well known, good explanation does not necessarily lead to good prediction (Kaplan, 1964, pp. 346-349). In discussing an example of a model below, we will illustrate several reasons why a good explanatory model may be a poor forecasting model.

Researchers in the natural sciences often aim to identify various "lawlike" relationships, not only to provide an apparatus for describing phenomena but also to capture and improve our understanding of hidden relationships in the real world. Thus the models of physics, for example, aspire to provide a sequence of ever deeper insights into the true nature of "reality". An improvement in the models then represents an advancement in science. New models, following the rejection of obsolete hypotheses that have proved inconsistent with experimental findings, are thus constantly altering the overall system of models which constitute the formal expression of our present understanding.

A necessary but not sufficient condition for the formulation of physical "laws" is the reproducibility of phenomena. Experimentally it is also important to be able to control "external" factors, so as to control the "laboratory setting". However, many areas of the natural sciences have to operate under less favourable conditions. A geologist, for example, cannot make big-scale controlled geophysical experiments of the development of the earth's crust.

The "lawlike" feature of natural science phenomena has made predictions possible to a sometimes surprising level of accuracy. One example is the prediction of the position of a certain planet on a given date in the future. Many people would argue that the precision of the prediction is due to a high level of understanding of the causal mechanisms of change which have been formalized in the model. However, there are aspects which might not be covered in the model which could invalidate the prediction. We cannot, for example, exclude the possibility of an unknown comet entering our solar system during the period of the prediction, thus altering its preconditions.

Let us compare studies of natural science phenomena with studies of the future development of social systems. In the case of complex social systems of the type often dealt with in futures studies, there is generally no reproducibility of phenomena and the possibility of controlling external factors is extremely limited. This complicates the development of explanatory models, as well as the testing of their applicability for forecasting purposes. However, a more important difference in this context is the dependence of the future development of social systems on human decision-making, and sometimes also on "new" causes of change. By "new" we mean here causes that have not existed in the past or that have existed but have not been very important. Thus, the step between a model reflecting some understanding of the present system and a model useful for forecasting is greater in the case of complex social systems than in the case of phenomena studied in the natural sciences. This is why we cannot really hope to develop explanatory models for complex social systems which can serve at the same time as long-term forecasting models.

In a paper entitled "Models in Forecasting and Analysis" Hermerén (1977) has distinguished between <u>mathematical</u> models and theoretical models. He argues that a theoretical model involves a theory about reality, i.e. a theory about some process or phenomenon based on the existence of laws or lawlike statements. This type of model thus involves a non-neglible amount of "understanding" or "explanation" as its bases ("explanatory model"). It is not unusual to count also theoretical models, when expressed in mathematical form, as mathematical models, but Hermerén wants to reserve the term mathematical model for non-theoretical models. A mathematical non-theoretical model may be a purely descriptive model or a model based on common-sense ideas without being based on a fully developed theory.

The models in mathematical form which are used in futures studies usually include relationships which have been formulated on the basis of "common-sense ideas". Although they may also include some elements of theory, they cannot be regarded as explanatory models fully developed for forecasting purposes. Consequently, they can be characterized as mathematical non-theoretical models, and can thus be said to belong to a different category of models than the natural science models which are theoretical models. Often, in fact, mathematical models that are used in studies of the long-term development of complex social systems are not even intended to be used for forecasting; it is often more adequate to describe them as "what-if" models or "learning models" (see p. 38).

## An example of a mathematical model

It has been observed that the numbers of certain species of animals in certain areas tend to vary according to a cyclical pattern. Also, where one type of animal eats another, an interrelated cyclical evolution has sometimes been observed for both types of animal. A model explaining phenomena of this kind has been presented by Volterra in "Leçons sur la Théorie Mathematique de la Lutte pour la Vie" (Volterra, 1931). We will now deduce a model of this type, referring to an island on which a number of foxes and rabbits both live, and where the foxes hunt the rabbits. Assume that the change $\Delta$ Rab in the number of rabbits (Rab) equals the number of rabbits multiplied by a fertility rate $\mu$, but that the foxes cause a relative decrease that is proportional (B) both to the number of foxes (Fox) and to the number of rabbits. This assumption can be written

$$\frac{\Delta Rab}{\Delta t} = (Rab) \cdot [\mu - B \cdot (Fox)]$$

Assume further that the rabbits are a very important source of food for the foxes, so that the number of foxes will tend to decrease when there are fewer rabbits. More specifically let us assume that during a time element $\Delta t$, the decrease in the number of foxes, if there were no rabbits, would be $\lambda \cdot$ (Fox). If rabbits were present, and more food were thus available, we could expect this decrease to be offset by a number proportional to the number of encounters between the two types of animal, i.e. $A \cdot (Fox) \cdot (Rab) \cdot \Delta t$ (Volterra, 1931, p. 14). These assumptions can be written

$$\frac{\Delta(Fox)}{\Delta t} = (Fox) \cdot [-\lambda + A(Rab)]$$

Our assumptions can thus be written as a system of differential equations, the solution to which constitutes the mathematical model. The type of solution obtained is illustrated on the next page. It displays the same type of cyclical evolution that had been observed. Hence, under certain circumstances, the model can be regarded as an explanatory model that reflects some understanding of the underlying causes of change. But how useful is it for forecasting?

Assume that we have actually been able to count the numbers of foxes and rabbits repeatedly over a certain period of time. This data can then be used to determine the coefficients in the mathematical model. Let us also assume that the model obtained fits all the observed data well, and perhaps also that we have used the model successfully for forecasting over a short period of time. But in a longer term, the two animal populations may deviate from the predictions of the model for a variety of reasons. Let us give some examples:

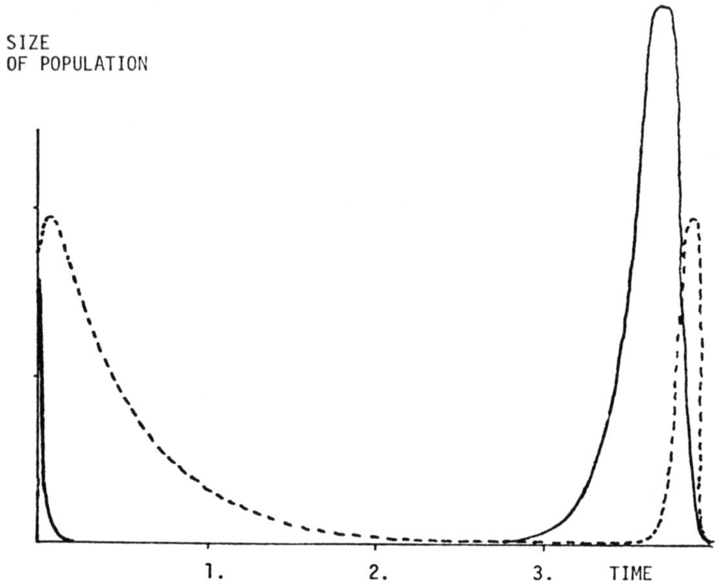


Fig. 3:2. Example of change over time of populations of foxes and rabbits.

a) Increase in influence of secondary causes of change

Obviously there are a number of "secondary" factors which might affect the change in the animal populations. We may not have included them in our model because we have been unaware of them or because we have not wanted to complicate our model unnecessarily, or because we have not had sufficient data to justify their inclusion.

Our model will still function well provided the excluded factors are small or "constant" (i.e. hidden in some of the constants of the model). But the importance of these secondary factors may increase. Assume for instance that the availability of food for rabbits depends to some extent on the weather, but that there were no serious problems during the observation period. However, there may be some slow long-term changes of climate, or there may be not exactly a change in climate, but certain "unusual" weather variations which nonetheless make a significant difference. Obviously our model may prove quite useless as a forecasting model during a period that includes changes of this kind.

b) New causes of change

In our model we assumed that our animals were living on an island, and we had thus implicitly fixed boundary conditions. But assume that the lake surrounding the island freezes, or something else happens that makes it possible for the animals to leave the island or for other animals to join them there. Or assume that there is a new and serious disease, perhaps more serious for the rabbits than for the foxes. Obviously our model will be quite useless for forecasting under conditions of this kind.

c) Human intervention

Obviously, too, the evolution of the animal populations may deviate in many ways from an earlier pattern because of human intervention. Assume for instance that the island acquires a new owner who has his own definite ideas about the kind of development he wants; perhaps he does not not want more than a certain number of foxes (or rabbits). If he allows hunting whenever this level is exceeded, the situation may well develop in a way that is quite different from that forecast by our model.

d) Approximations

As is well known, there are various fairly trivial reasons why a model may give quite misleading results when it is used for long-term forecasting. The input data may involve approximations. Sometimes errors propagate through the solution procedure and may become very troublesome. An illuminating example of this mechanism is provided by the difference in results between the following two systems of equations:

$$\left.\begin{array}{l} x - y = 1 \\ x - 0{,}99y = 1{,}01 \end{array}\right\} \quad \begin{array}{l}\text{giving as} \\ \text{result}\end{array} \quad \begin{array}{l} x = 2 \\ y = 1 \end{array}$$

and

$$\left.\begin{array}{l} x - y = 1 \\ x - 1{,}01y = 1{,}02 \end{array}\right\} \quad \begin{array}{l}\text{giving as} \\ \text{result}\end{array} \quad \begin{array}{l} x = -1 \\ y = -2 \end{array}$$

In our example we had a system of differential equations. Such systems are often solved by computers and approximate procedures are used for obtaining the solutions. This type of approximation, too, may cause errors in the solutions.

Our example illustrates that models cannot easily be made to provide accurate forecasts in the long run, even when we are dealing with fairly simple systems. But the models may have other uses. For instance they may sometimes be used to examine the possible differences between decision alternatives in terms of future impacts. A model may reflect such differences fairly accurately, even if it is poor at forecasting. A model may also indicate how the system would develop under a number of different assumptions. By using such model we thus learn about the behaviour of the system (under simplified conditions). This is why models in futures studies are sometimes called what-if models or learning models. Their role is thus to help the user to gain general insights, perhaps to indicate the possible consequences of certain actions, given certain assumptions. Operations research literature often deals with normative models for the development of a system, usually over a short period of time. Such models are designed for studies of possibly desirable changes, given certain objectives, i.e. they can be regarded as a type of what-if-model.

The expression "learning model" seems to have come into general use in connection with "operational gaming" in the 1950s and 1960s (cf. Thomas, 1961). One of the uses of this technique is in the study of moves and countermoves in war games or business games. Even when the outcome of such games is determined in quantitative terms, the main purpose is not usually to find the outcome of the simulated chain of events but to enable the participants to acquire some understanding of the problem area concerned. Similarly, when a mathematical model yields a reference projection or an estimate of what could happen in the future under a certain set of assumptions, it adds to our understanding of the problem concerned.

To help us to formulate mathematically assumed interdependencies in a complex system, various symbolic languages have been developed. Two symbolic languages, those of Forrester and Odum, are the best known. In Forrester's language the "foxes and rabbits" problem could be expressed as follows:

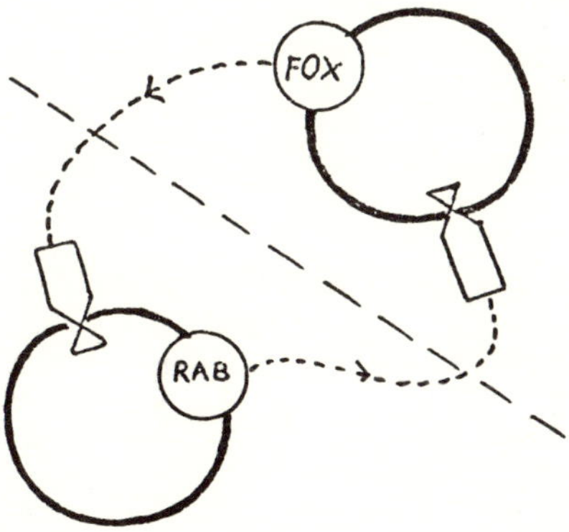


The illustration provides a shorthand statement of the following observations:

- there is one population of foxes and one of rabbits (two circles respectively)
- the number of foxes tells us how many new foxes (or how many fall-offs) are to be added to the populations of foxes (and the same for rabbits) - wide circular arrow

- however, this input of new individuals (or losses of individuals) in the population of foxes is controlled by the number of rabbits (dotted line to the controlling point on the circular arrow) (and the same for rabbits)

It is easy to see that this symbolic notation provides a useful tool in the modelling of systems more intricate and complex than that of the "foxes and rabbits". Incidentally, the world model developed by Meadows (based on Forrester's model) is of the same type as the one illustrated in our example, i.e. it consists of set of differential equations, and as an aid in its formulation Forrester's symbolic language was used (Meadows et al, 1972). The strong international criticism of this project has partly been directed against the model itself and partly against the way in which conclusions were drawn from the computer runs (cf. Cole, 1977; Schwarz, 1974). Since we regard the model as a mathematical "what-if" model and not as a forecasting or theoretical model, our main criticism is that the authors overestimate the possibility of drawing definite conclusions from models of this type. We will return to this problem in Chapter 5 when we discuss the use of mathematical models in our case studies.

## Different types of mathematical models in futures studies

In models of long-term developments, the model-builder's assumptions about the causes and mechanisms of change are of fundamental importance. In the fox and rabbit example discussed above, involving a system of differential equations with coefficients derived from historical data, there is an implicit assumption when the model is being used for forecasting that the causes and mechanisms of change will remain the same in the future as in the past - or rather, not exactly as they have been in the past but as the modeller has modelled his perception of the past. However, there are other types of mathematical models which are quite different in this respect. We could distinguish between mathematical models in which the mechanism of change in the future

a) is wholly determined internally, on the basis of past behaviour

b) is wholly or partially determined exogenously

c) remains outside the scope of the problem treated by the model

Examples of type c would be models of the way in which new systems, perhaps a new aeroplane or a transportation system, will function in the future. Models of this kind are concerned with what would happen if such a new system were developed and put into operation.

Meadow's world model belongs to category a. An example of a model of type b is "The UN World Model", a model of the input-output type developed by W. Leontief (Leontief, 1977). Here assumptions about future economic growth rates and population growth are made exogenously. Leontief has characterized this model as a "planning device". Another example of a type b model is MARKAL, the energy system model developed by OECD's energy agency, IEA (Abilock & Fishbone, 1980). In the description of the model it is emphasized that MARKAL is intended not for forecasting but for examining what difference it would make to the future development of the energy system if certain new energy technologies were to become available.

In conclusion we would suggest that mathematical models in futures studies are not usually intended to yield predictions, although such a view may not be generally accepted. Hermerén (1977), for instance, seems to assume implicitly that the purpose of these models is prediction. In fact mathematical models are often presented and used in futures studies as "planning devices", "what-if models" or "learning models". We feel that these three concepts largely overlap, although there are slight differences in emphasis. The first two put the emphasis on using the model for comparisons of different decision alternatives or hypotheses. As we have already mentioned, a model may give a reasonably good estimate of such differences even if it is inadequate for forecasting future consequences in absolute terms. The term learning model emphasizes the use of the model to gain general insights into the problem area being studied, i.e. the soft use of a hard tool.

In a comparative evaluation of global models, Richardson has illustrated how in many instances the models have been influenced as much by the modellers' background as by the clients' needs (Richardson, 1978). Obviously, such an impress on a model may make it difficult and even dangerous to draw definite conclusions from specific outcomes of model runs. Richardson considers such general conclusions as can result from the experience of developing and using a model over a long period of time to be perhaps the most valuable kind. Various studies of global models also show that a comparison of assumptions and results from a set of different model projects can provide important insights. The importance in policy studies in general of simultaneously using several different assumptions and models has been emphasized (Dror, 1974, Mitroff & Pondy, 1974). The possibility of combining the use of a mathematical model with other approaches in a futures study should also be mentioned (cf. Section 5.4).

## REFERENCES

ABILOCK, H. & L. FISHBONE, User's Guide for MARKAL (BNL Version), A Multiperiod, Linear Programming Model for Energy Systems Analysis, 1979 (IEA Energy Systems Analysis Project, Brookhaven National Laboratory, Upton, New York 11973, U.S.A.).

ANDREN, N. & B. SCHWARZ, Framtidsvisioner i långsiktig planering (Centralförbundet Folk och Försvar, Stockholm, 1974).

BOX, G.E.P., and G.M. JENKINS, Time Series Analysis, Forecasting and Control (San Francisco, Holden-Day, 1970).

BROWN, S., "Scenarios in Systems Analysis" in E.S. Quade and W.I. Boucher (eds), Systems analysis and policy planning applications in defence (American Elsevier, New York, 1968).

CALDER, N., "What is Futures Research?", *New Scientist*, October 1967, pp 354-55.

COATES, J.F., "Review of Sackman Report", *Technological Forecasting and Social Change*, Volume 7 (1975), Number 2, pp 193-94.

COLE, S., Global Models and the International Economic Order (Pergamon Press, Oxford, 1977).

DROR, Y., "Some Features of a Meta-Model for Policy Studies", *Policy Studies Journal*, Volume 3 (1975), pp. 247-255.

EDMAN, M., "There is no Naive Way of Making Forecasts" in Knowledge and Concepts in Futures Studies, S. Schwarz (ed) (Westview Press, Boulder, Colorado, 1976).

GODET, M., The Crisis in Forecasting and the Emergence of the "Prospective" Approach - With Case Studies in Energy and Air Transport (Pergamon Policy Studies, Pergamon Press, New York, 1979).

GORDON, T.J., and O. HELMER, Report on a Long-Range Forecasting Study (P-2982, Sept 1964, Rand).

GÄRDENFORS, P., "On the Information Provided by Forecasting Models", *Technological Forecasting and Social Change* 16 (1980), pp. 351-61.

HERMEREN, G., "Models in Forecasting and Analysis", in Trends in Planning, Edited by C-G Jennergren, S. Schwarz, O. Alvfeldt (FOA, Stockholm; Wiley, London, 1977).

HILL, K.Q., "Trend Extrapolation" in Handbook of Futures Research, J. Fowles (ed) (Greenwood Press, Westport, Connecticut, 1978).

de JOUVENEL, B., The Art of Conjecture (Basic Books, New York, 1967).

KAHN, H. & A. WIENER, The Year 2000 - A Framework for Speculation on the Next Thirty-Three Years (MacMillan, New York, 1967).

KAPLAN, A., The Conduct of Inquiry (Chandler, San Francisco, 1964).

LEONTIEF, W. et al, The Future of the World Economy (United Nations, New York, 1977).

LINSTONE, H. & TUROFF, M. (eds), The Delphi Method - Techniques and Applications (Addison - Wesley Publishing Company, Reading, Massachusetts, 1975).

MARTINO, J., Technological Forecasting for Decision making (American Elsevier, New York, 1972).

MEADOWS, D., et al, The Limits to Growth (Universe Books, New York, 1972).

MITROFF, I. & L. PONDY, "On the Organization of Inquiry: A Comparison of Some Radically Different Approaches to Policy Analysis", *Public Administration Review*, XXXIV (Sept/Oct, 1974, pp. 471-79).

NANUS, B., WOOTEN, L.M. and BORKO, H., The Social Implications of the Use of Computers across National Boundaries (AFIPS Press, Montvale, New Jersey, 1973).

NANUS, B., "Interdisciplinary Policy Analysis in Economic Forecasting", *Technological Forecasting and Social Change* 13, pp. 285-95 (1979).

RIVETT, B.H.P., "Futures Literature and Futures Forecasting - A Critical Review", *OMEGA*, The Int. Jl of Mgmt Sci., Vol. 7, No. 1. pp 33-41.

SACKMAN, H., Delphi Assessment: Expert Opinion, Forecasting, and Group Process (R-1283-PR, April 1974, Rand).

SCHWARZ, S., "Apokalyps och karisma: Tankar kring boken tillväxtens gränser och fenomenet Rombklubben" in Än se'n då - diskussioner om framtidsstudier (Föreningen för Framtidsstudier och Ingenjörsförlaget AB, Stockholm, 1974).

THOMAS, C.J., "Military Gaming", Ch. 10 in Progress in Operations Research, Vol. 1 (John Wiley & Sons, Inc. New York, 1961).

VOLTERRA, V., Lecons sur la Théorie Mathématique de la Lutte pour la Vie (Gauthier-Villars, Paris, 1931).

WIKSTRÖM, S., Våra dagligvarors distribution (Liber, Stockholm, 1974).

WILDHORN, S., How to Save Gasoline (Rand, 1976).

WILSON, I.H., "Scenarios" in J. Fowles (ed) Handbook of Futures Research (Greenwood Press, Westport, Connecticut, 1978).

# 4
# Three Studies of the Future: Swedish Experiences of Futures Studies and Policy-Making

We have already referred to the large and growing literature concerned with individual techniques of futures studies and with various methodological experiences from some large-scale modelling undertakings. We have also commented on the apparent lack of concern with the methodological problems engendered by the great number of futures studies which all - despite their variety in other respects - have policy problems on a national level as their point of departure. Our own discussion, although geared to issues that are familiar across a wide range of futures studies, takes three such "mid-level" projects as its source of examples and references.

Each of these three studies was launched by a policy-making or planning body at the national level in a particular country, namely Sweden. They were conducted during roughly the same period, the late 1970s. And they were all characterized by an ambition to make an impact on policy-making and on the public debate. Furthermore, we have had ready access to all material relevant to the three studies. In two cases - Resources and Raw Materials, and Transportation - two of the present authors were directly engaged on the projects. The third author kept in continual and close touch with the Resources and Raw Materials project and with the Energy and Society study. Interviews were held with members of the project groups and with the leading politicians and civil servants involved; official documents and internal memoranda from these and other project groups at the Secretariat for Futures Studies were examined and analysed.

Despite their similarities we feel that the three projects are sufficiently distinct both in their general approach and in their contributions to policy-making and political controversy, to justify a fairly detailed presentation.

The projects Energy and Society and Resources and Raw Materials were two of the four first-generation projects launched by the Swedish Secretariat for Futures Studies in 1974. The Secretariat had been established in 1973 on the recommendation of a government commission whose report "To Choose a Future" (1972) included a review of the field of futures studies in Sweden and abroad (cf. Wittrock 1979 and 1981).

A salient political motive for setting up the commission had been the desire to evolve a conception of futures studies in opposition to the one that lay behind "most futures studies produced so far ... sponsored by military establishments and by the world's major multinational corporations" to quote Alva Myrdal who was chairman of the commission and, at the time, a member of the Swedish Cabinet. It was feared that such "studies may be based on scales of values that are not democratically acceptable. We must avoid any 'colonizing of the future' by powerful interest groups, national or international".

At first the Secretariat was attached to the Prime Minister's Office but in 1975 the Riksdag decided, in accordance with a proposal of the conservative, liberal and centre parties, that the Secretariat should be separated from the Social Democratic Prime Minister's Office and transferred to the Ministry of Education. This change undoubtedly restricted the ability of the Secretariat to act as an agent for improving the cross-ministerial and long-range planning base for the Ministries' actions and decisions. But the effect on the current work of the Secretariat - e.g. on the four projects on Energy and Society, Resources and Raw Materials, Sweden in World Society and Working Life in the Future - that were already under way, was slight. The themes of these projects had been selected after discussions with representatives of various ministries and an analysis of the comments from some 130 government agencies and non-government organizations on the report "To Choose a Future".

The project groups had 4-6 full-time members. They also made use of at least 20 other people who contributed to reports in an expert capacity and on a part-time basis. The groups were independent vis-à-vis the government, but each one had an inter-ministerial reference group, headed by an under-secretary of state, which maintained continual contact with the group and commented upon its work.

The project groups' plans and reports could be brought up at the meetings of these reference groups, and they were the subject of lively debate on several occasions. It was not the job of the inter-ministerial groups, however, to approve project reports for publication. This system permitted government officials to be involved in the work of the projects, while also allowing the project groups to adopt a very open and sometimes provocative style of reporting.

Contrary to the Energy and Society and the Resources and Raw-Materials studies the Transportation Study was not initiated by the Secretariat of Futures Studies but by an agency of the Ministry of Transport, the Transport Research Delegation. From the guidelines given for this project at its start it is clear that the underlying ideas of its initiation stemmed from a long-range planning rather than a futures study tradition. But the project was similar in scope to the projects of the Secretariat of Futures Studies. Two of the members of its reference group were former members of the Alva Myrdal Commission. The Transportation Study was initiated in 1975 and its report, "Transportation and Transport Research - Future Directions", was published in 1978 (Schwarz & Svensson, 1978).

## 4.1 ENERGY AND SOCIETY

### Policy background

The project Energy and Society was one of the four first-generation projects of the Swedish Secretariat for Futures Studies. The launching of a futures study on energy policy in 1974 did not mean that no long-range studies had been made in this field before. In fact, because of the long lead times in introducing new technical systems, energy policy had provided fertile ground for the evolution of a forecasting tradition, primarily in connection with electrical energy.

In the Swedish administrative and planning system of the early and mid-1970s, energy forecasting was undertaken by several public bodies: the Energy Forecasting Commission outlined alternatives for the periods 1975-1985 and 1985-2000 and the Energy Research and Development Commission had called for studies covering the period up to the year 2000. The National Industrial Board and the Central Operating Management, an organization for co-operation between the main private and public Swedish power producers, are other bodies which have been engaged in energy forecasting activities.

At the ministerial level the major resposibility for energy policy planning lay with the Ministry of Industry. But other ministries, e.g. the Ministry of Housing and the Ministry of Transportation were also directly involved in energy policy-making. By and large, contact between energy forecasters and planners connected with a variety of bodies and commissions was well developed when the energy futures study was started.

The oil crisis of 1973-74 dramatically illustrated the vulnerability of Sweden's energy supply and that of most other Western countries. There no longer appeared to be any guarantee of a steady supply of inexpensive fossil fuels. And Sweden's dependence on imported oil for some 70 % of its total energy supply suddenly became an urgent problem. Energy policy, it was generally argued, should aim at a reduction in oil imports and the promotion of domestic energy sources. Admittedly, the idea of replacing foreign energy sources by domestic ones was not new in the Swedish context. Similar considerations lay behind the major technology programme launched in 1956 which had aimed at the development and introduction of heavy-water nuclear technology. In view of Sweden's possession of substantial uranium deposits, this programme was believed in the late 1950s and early 1960s to hold the promise that, in a long-range perspective, the nuclear fuel cycle could be kept within the country; and when at some time in the future imports of fossil fuels began to dwindle, domestic nuclear power would supply the energy needed to keep industry expanding and welfare growing. By 1970, however, this massive heavy-water nuclear technology programme was officially abandoned, and long before that the thrust of the programme had already been curtailed. Rising costs, unenvisaged problems at the demonstration stage and the early commercial competitiveness of American and domestic light-water reactors spelt doom for the "Swedish line" in heavy-water nuclear technology. However, by the early 1970s an elaborate informal network had evolved embracing researchers, government officials and representatives of industry who tried to promote research and development in the energy field. Ideas and requirements were also formulated, recognizing the need for a wide range of public measures designed to solve long-range energy supply problems, and in this way it was possible, for example, to influence the work of a government commission, namely the Energy Programme Commission. This commission presented a report in 1974 which provided the foundations for research and development on a large scale in the energy field. The programme was launched in 1975.

As a result of the new awareness of the urgent need for a coherent long-term energy policy, a special Energy Policy Delegation - a co-ordinating body at the under-secretary of state level - was set up early in 1974, and work started in the government offices on drafting a comprehensive Energy Policy Bill to be presented to the Riksdag in 1975. In the spring of 1974 this delegation and the Secretariat for Futures Studies jointly launched a project on long-range options in Swedish energy policy. The idea was that this futures study project on Energy and Society should try to identify consequences over a far longer period than was customary in the forecasts produced by such bodies as the Energy Forecasting Commission, the National Industrial Board or the Central Operating Management.

The life-time of the Energy and Society project, 1974-78, coincided with a change in the political importance of energy policy in general and nuclear power in particular.

Up to 1973 decisions regarding nuclear power had not involved any real cleavages between the political parties. From 1973 and throughout the remainder of the 1970s nuclear power policy became a major issue on the political agenda. This was due not least to the strong anti-nuclear stance in the early and mid-1970s of the leading non-socialist party, the Centre Party.
The issue of nuclear power was of crucial importance both to the outcome of the 1976 general elections and the concomitant change from a Social Democratic to a non-Socialist government, and to the break-up of the new three-party coalition cabinet in October 1978 (Peterson 1977; Holmberg, Westerståhl & Branzén 1977; Peterson 1979; Vedung 1979).

## Problem formulation

A starting-point for much of the discussion in the Western world and also for the work of the Energy and Society project group, was the above-mentioned assumption that continued growth in energy consumption could no longer be combined with a predominating reliance on fossil fuels. If this assumption held, it seemed natural to focus a study of energy futures on the determinants and characteristics of various levels of energy consumption. This was essentially the strategy chosen by the Ford Foundation's study of American energy policy. In their report "A Time to Choose" three main development alternatives were discerned: historical growth, technical fix, and zero-energy growth, involving 3.5 %, 2 %, and (from 1990) 0 % annual growth in energy consumption respectively. For each of these alternatives it was possible to examine the effects on economic development, employment, capital requirements, environmental problems, and on the orientation of the production of various goods and services.

Originally, the Swedish study adopted a similar approach, although a slightly different set of characteristics was to be analysed, including effects on working environment, income distribution, and the developing countries. The group hoped thereby to be able to survey energy problems from a wide range of viewpoints or, to use the group's own words, through different windows. And it was hoped that the outlines of a comprehensive energy policy would emerge from a series of onslaughts from different angles.

However, the group gradually abandoned its focus on levels of energy consumption and problems connected with the physical scarcity of energy resources. Instead it came to look upon energy problems in terms of the transition over time of combinations of energy technologies based on different energy sources.

The discussions in the group that were instrumental in bringing about this conceptual change were connected with the enlargement of the project group, whereby the core group came to include five full-time members. Apart from the project leader - who was an engineer by training, with a background of administrative service at the National Audit Bureau - there was one systems analyst, one nuclear physicist who was associated among other things with the Environmental Studies Programme at the University of Lund, and two economists. The discussions in the group revealed major differences of opinion, particularly between the economists and the other members. The formulation of the problem was a lengthy process involving much adjustment and clarification and lasting well over a year.

The experience of the energy project suggests that a definition and conception of a futures problem which is sufficiently explicit to provide the project with a structure but open-ended enough to permit a wide range of empirical material from different areas of knowledge, is likely to be the most fruitful. During the process of adjustment and clarification, and in fact throughout the project, the inter-ministerial reference group was used mainly as a source of expertise and feedback often on matters of a specialized or technical nature, but it was never allowed to exert any substantial influence on the general course chosen by the project group. In fact only one meeting attended by the whole reference group ever took place. This was in marked contrast to the practice of the Resources and Raw Materials project group, which established a pattern of regular contacts and meetings with its reference group.

In the case of the Swedish energy futures study the basic problem of energy policy was understood to be the management of a transition from dependence on oil to reliance on less scarce - but perhaps more costly - energy sources. But, analytically speaking, such a conception is open to many different interpretations.

The American economist William D. Nordhaus's article "The Allocation of Energy Resources" might serve as one example. Nordhaus has outlined a schema for the allocation of global energy resources for a series of phases starting with five ten-year periods, followed by two twenty-five periods and two fifty-year periods. The final stage involves the transition to breeder technology, which would gradually be introduced from 2020 and which would have completely replaced fossil fuels and other energy sources by 2120. Transitions over time in the utilization of energy technologies are estimated so as to minimize the costs of meeting a set of demands. The main determinants in this analysis are the costs of production.

In the Energy and Society project a rather different conception of transition was formulated. The group did not see the introduction of energy technologies based on different energy sources mainly as a matter of adaptation to changing costs of production. Instead the group regarded such changes as highly dependent upon technical and organizational systems that can be affected by acts of political will. Such differing interpretations of the problem of transition suggest that fairly fundamental ideas about the nature of society and of man's place in society - e.g. the possibility of consciously changing a given state of affairs - are relevant to the theoretical assumptions on which futures studies are based.

## Two energy futures

To clarify the range of options open in a long-term perspective the Energy and Society group outlined two extreme alternatives for the Swedish energy system in the year 2015. These alternatives can be regarded to some extent as scenarios, but they were not so designated by the group itself. When in its final report, "Solar Versus Nuclear: Choosing Energy Futures", the group outlined a possible chain of events leading to one extreme alternative, "Nuclear Sweden", it used the term "intellectual experiment" to characterize its approach.

The choice of the two extreme alternatives was based on the argument that there are essentially three kinds of long-term energy sources: coal, uranium, and renewable energy sources. Coal, like oil, does not constitute a domestic Swedish source and could thus entail the risk of new types of dependency and vulnerability in the country's energy supply, if it were to replace oil on a major scale as an energy source in the Swedish context. Furthermore it shares with oil a number of negative environmental consequences. The two extreme options that were examined in some detail were termed "Nuclear Sweden" and "Solar Sweden" respectively, two catch-words that proved to have a wide appeal not least during the nuclear referendum campaign in the spring of 1980.

According to the solar alternative, fossil and nuclear fuels would have been entirely replaced by renewable energy sources by the year 2015. According to the other alternatives, nuclear energy would account for more than 80 % of the total energy supply. The remainder would come from hydropower and bark and lye in the same proportions to one another as in 1975.

The group then posed the question as to whether these two alternatives were in fact feasible. A starting-point for the outlines of the two extreme alternatives was the specification of a set of conditions common to both, namely:
time perspective - the year 2015; population of Sweden - same as today; number of dwellings - an increase of 40 % from 1975, with a slightly larger increase in the total space of dwellings; production of goods and services - an increase of 100 % between 1975 and 2015 (slightly less than an annual 2 % growth rate); specific energy consumption - a decrease between 1975 and 2015 of 20 % in industry, 50 % in the service and transport sectors, and of 30 % in dwellings.

The feasibility of the two extreme alternatives in the light of these conditions was then examined with respect to the following dimensions: technology, economy, siting and land-use planning, industry structure in the energy sector, capital market for the energy sector and requirements for qualified manpower in the energy production sector. The overall conclusion was that even with fairly conservative assumptions abouts costs and available technologies neither of the extreme options could at the time be declared unfeasible. Both seemed to be able to satisfy the set of conditions specified at the outset. But this did not mean of course that they might not prove undesirable or impracticable. But so much uncertainty surrounded both alternatives, that neither of them could be excluded at that point.

In order to retain freedom of action - a key concept in the analysis of the group and one that undoubtedly owes much to the tradition of adaptive planning - it was essential to formulate short- and medium-term policies of such a kind that neither of the extreme alternatives, nor any mixed alternative, was inadvertently precluded. Thus, to achieve this freedom, it seemed necessary to clarify the mechanisms that determine the choice of energy systems. Otherwise there would obviously always be the risk that future options were being unintentionally precluded.

In its analysis of potential threats to the maintenance of freedom of action, the group concluded that the crucial factors were mainly technical and organizational. The analysis hence dealt with problems concerning the amount and quality of energy used, the existence of large complex systems of energy production (e.g. the nuclear fuel cycle) and energy distribution, and problems of organizational inertia. In this last context it was argued, for instance, that the orientation of research and of existing administrative structures, rules and policies were adapted to already operating technologies. Such organizational factors affect the competitiveness of different technologies, for example through the workings of the capital market. Active measures would therefore be necessary in a short- and medium-term perspective to achieve a balance between the nuclear and the solar options, measures pertaining not only to research and development but also to the amounts and qualities of energy used and to the nature of the energy transmission and distribution systems. The sort of question to be considered would concern, for instance, the degree of centralization in the distribution system, the choice between electricity or water-based heating systems or a combination thereof, and so on.

Thus, the main driving-forces identified on a basis of various background studies etc (the evolution of Swedish energy policies in the period 1890-1970 was the subject of one such study) depended on technological and organizational rather than on economic, political, or other factors. International economic dependencies and domestic economic constraints were not regarded as precluding any of the options. The energy system in a broad sense - including not only energy production but also distribution systems and consumption patterns - was seen as subject in the last instance to control by public action.

Obviously, the project group did not try to predict what kind of energy system would actually be in use in Sweden in the year 2015. Instead the group emphasized the uncertainties that underlie any choice of energy future and stated that its work was "oriented towards the problem of avoiding bad solutions rather than trying to choose optimal ones today". Hence it was necessary to outline such chains of events as might indicate how certain sets of decisions could engender or prevent undesirable solutions. The group argued that the two extreme alternatives served to clarify the range of options currently available and to highlight dilemmas that might arise on the way towards them.

The ensuing public debate in Sweden on the "Solar Versus Nuclear" report suggests that the very notions of a nuclear or a solar Sweden were politically explosive in themselves. Whatever might have been hoped, it seems that the methodological subtlety of retaining freedom of action and regarding the two extreme future alternatives as theoretical constructs has a slim chance of being appreciated in the midst of a heated political discussion. However, there are several interesting methodological features of the Energy and Society project that have received comparatively little attention. One concerns the project group's decision not to make use of traditional futures-studies techniques such as Delphi nor to dwell too much on the technical problems of scenario construction; another concerns the role of extreme alternatives as opposed to mixed alternatives in outlining a range of future options; a third is connected with the balance in the study between the presentation of alternative futures on the one hand and the analysis of determinants and key variables that might affect the likelihood of the different alternatives on the other. This last point led to a lively debate within the project group itself when it came to writing the final report.

## Implications and impacts

In a turbulent environment - where energy policy has been a major source of political conflict - the Energy and Society group advocated a short- and medium-term energy policy that might serve to keep major long-range energy policy options open. This thrust of the group's argument was partly a consequence, as we have seen, of the professional inclination of some of the group's members. But it also seems to have been connected with differences of opinion within the group itself as to the desirability of various energy policy alternatives and the appropriateness of expressing such possible preferences in the group's reports. Nevertheless the group was able to act as a coherent unit, while permitting individual members substantial freedom of action outside the scope of the group proper. Thus the project leader could contribute to the Social Democratic debate on energy policy, for example by writing articles together with the chairman of the project's reference group, who was an energy policy advisor to the former Social Democratic government. The group's stance was also perfectly compatible with the fact that another of its members acted later as energy policy advisor to the Centre Party energy minister in the subsequent three-party coalition government. However, since political actors in general either did not share the group's passion for non-commitment, or did not regard it as entailing the consequences for short-term policy suggested, the group's advocacy of non-commitment could not be other than problematic.

The project group's analysis had obvious policy implications in both the short- and medium-term which, from the point of view of policy planners and administrators in the ministries and the State Power Board, were far from orthodox. Nonetheless, the views of the project group have received widespread attention in the public debate in Sweden, and its main reports have been published in English and have attracted international attention. The group also succeeded in transferring some key elements of its analysis to government commissions such as the Energy Commission of 1976-78 and the Energy Research and Development Commission. However, the use made of the results from the futures study, both by the Energy Commission - on which individual members of the futures study project were active in an expert capacity - and by various politicians and groups, seems to have been somewhat selective. Certain findings, illustrations and key concepts have been adopted, but it is still not clear how far the project group's insistence on the necessity of understanding the driving forces behind the emergence of energy systems has been appreciated. The change in the political environment from a "closed" policy-making environment on the energy front to one of redefinition and turbulence, was probably necessary before an unconventional analysis such as that of the Energy and Society project group, could have had the impact that it did. Indeed, in a different climate the study might never have got off the ground in the first place, and there was in fact some initial resistance on the part of the Ministry of Industry to the idea of a futures study project dealing with the energy issue. And in 1975, as a result of protests from the Social Democratic under-secretaries of state of the Ministries of Industry, Commerce and Finance against the publication of the project's introductory "Energy and Society" report, the ties with the Energy Policy Delegation were severed and the project became responsible solely to the Secretariat for Futures Studies. Developments since the publication of "Solar Versus "Nuclear" have also had implications for the relevance of some of the group's findings. In particular, the referendum on nuclear power in the spring of 1980 showed - and later declarations by leading politicians have confirmed - that an overwhelming majority of the Swedish people want a long-range Swedish energy future that does not include nuclear power. Thus, neither the scenario implying a nuclear Sweden by the year 2015, nor the building of any nuclear power plants over and above the 12 already in operation or under construction, can hardly be regarded any longer as feasible political options. The extent to which the activities of the energy futures study project has influenced this development or indeed determined the range of options considered in the referendum (none of which corresponded to a predominantly nuclear Sweden) remains an open and fascinating question.

## 4.2 RESOURCES AND RAW MATERIALS

What can be regarded as methodological questions in a multidisciplinary undertaking on a theme such as "the future of raw materials"? Perhaps we can distinguish three separate phases in the early life of such a project, each with its own kind of problems.

| | | |
|---|---|---|
| 1. | The decision in an organization to initiate a project. | Here the important issues are the choice of theme in relation to the institutional context, the formulation of the problem, the appointment of members to the project group and the assignation of their task. |
| 2. | The formation of the project group and first interpretation of its task. | Here the methodological issues concern the actual composition of the group and its links to reference groups and consultants. The interpretation by the project group of the task and the choice of general approach can be of decisive importance. |
| 3. | The choice of methods and tools, including subsequent re-evaluation in the course of the study. | Here questions arise about the proper use and mix of such methods as scenarios, Delphi, econometric models etc. |

### Phase 1

Like Energy and Society the Resources and Raw Materials project was attached institutionally to the Swedish Secretariat for Futures Studies presented above, and belonged to the first generation of the Secretariat's studies. This meant that the study was to address problems that appeared likely (sooner or later) to be important to government decision-making. However, decision-makers in the Riksdag (the Swedish Parliament) and the Government were not expected to be the only potential customers of the results of the study. The study also addressed the people at large, and in fact a basic idea underlying the study was that it should promote an active public debate as a basis for a viable democracy. (The price of subscribing to the reports was fixed more or less to cover the cost of postage!)

The choice of the particular theme of Resources and Raw Materials for one of the projects of the Secretariat of Futures Studies was based on various ideas about the importance of these problems. In an early document produced by the Secretariat in which the scope of the study was outlined the following arguments were put forward to justify the choice of this theme:

- intensive international discussion and national studies already published elsewhere
- rising prices of commodities 1972-74
- limited stocks
- producer cartels
- anxiety about long-term shortages
- the "Limits to growth" report
- discussion in connection with the UN Population Conference

The project group was asked:

- to evaluate existing international raw materials/resources studies
- to analyse the structure of Swedish dependence on various raw materials.

Apart from these requirements the scope of the study remained fairly open. However, one specification was that it should focus on Swedish conditions. "The international development as outlined here will, however, provide an important background for the study".

The study was not meant to result in a forecast of any inevitable future development. Instead the idea was to outline alternative paths of development and situations and to examine the question of when government intervention might promote or counteract various types of developments.

The study was to cover a broader time span than is normal in sectoral planning studies.

The directions given at the outset also indicated areas of interest for special studies e.g.

- trade policy
- links between the supply of raw materials and societal problems, such as employment
- recycling.

<u>Phase 2</u>

The members of the project group represented a variety of academic and other backgrounds; on the academic side economics, economic history, sociology and physics were all represented. For a limited time the team also included an ecologist. The group also embodied experience of working in government administration (particularly foreign aid, economic policy) and experience of technology assessment.

A special reference group was established, whose main task was to facilitate contacts between the project group and various ministries and government units. The reference group was headed by the Under Secretary of State for the Ministry of Commerce.

After consultation with the reference group, the project group developed a set of priority areas and started to build up a network of consultants, drawn mainly from academic circles but also from industry and government.

It is interesting to look back to see how the project group regarded its task in the initial stages of its work. In a memorandum written after roughly 3 months, several basic issues were settled on the question of a suitable approach.

It was felt to be important that the group should define a set of values to serve as guidelines for the study. Some guiding values had already been given by the Secretariat at the outset, and these can be summarized as follows:

- The future is not deterministically given - it can be chosen.
- A democratic society is an overriding aim and this requires well-informed participating members.
- It is important to make studies on a national basis, as value judgements may have a decisive influence on the outcome of the studies. Thus the images of the future should not be formulated solely by commercially oriented internationally operating institutions. In the presentation of the studies, value assumptions should be stated explicitly. Alternatives should be presented. ("It should be possible to establish how different values could be fulfilled in different alternatives").

- Man is regarded as a basically active being with an ability and a desire to create and to assume responsibility.
- International solidarity should be considered as an essential goal.
- The concept of justice is to be applied on the international as well as the national level, between different social groups of people, between different ages, between men and women.
- The choice of methods of production and the use of resources should take into account the rights of future generations.

In addition to these value judgements the project group noted:

- That more active planning is needed in the use of resources, since price formation in a free market does not take external effects into account.
- That such planning should be based on a holistic view of the use of resources, involving in particular an integration of economic and natural-science ways of thinking.
- That the distribution between rich and poor countries reflects an unfair international system that needs changing. But an "international perspective of justice" cannot be treated independently of Swedish economic equity issues.

The feeling in the project group in general was that on the resources/raw material front the study should provide an overview of future developments in Sweden against an international background. Two time-frames were chosen: 10 years and 25 years.

Phase 3

The work was to include the following steps:

- the formulation of a set of values

- the initiation of background studies surveying available data and facts
- the development of outlines of scenarios
- the development of final versions of the scenarios based on the new facts made available in the background studies.

The scenarios were thus thought to be of central importance to the study, both as a tool of investigation and for the presentation of the final results.

It might be of interest to note that "scenarios" at this early stage of the project were vaguely regarded as a set of descriptive "essays" illuminating different types of future developments and situations.

The logic of the project, as envisaged by the project group at this stage, is outlined in Figure 4:1.

It is interesting to observe in retrospect that the scenarios <u>did not</u> come to play the central role that was anticipated at the outset. This was due not to lack of effort but to a growing feeling in the project group that the "results" were "too soft". They were felt to lack the power to support their own validity. In particular the problem of internal inconsistency was found to be very troublesome. But more of that below.

When the study had been completed it was found that instead of following the project path presented in Figure 4:1 work had tended to proceed by developing three areas <u>in parallel</u>:

I - Background studies

II - Scenario work

III - Studies using econometric models

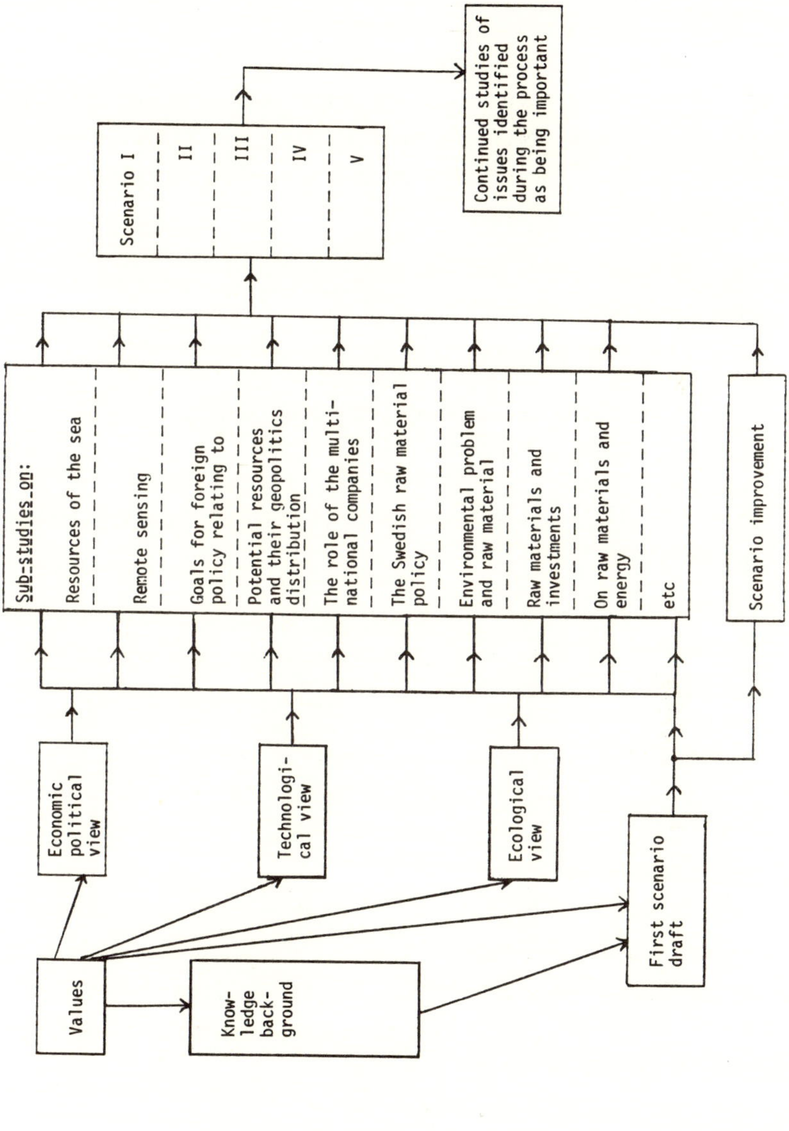


Figure 4:1 Original plan for the project

followed by

IV - Work on a synthesis.

Before returning in more detail to these four distinct modes of operating we should briefly state the major deviations between the actual path of the project as compared with the early plans as outlined in Figure 1:

a) The different parts of the study have been given a more autonomous role in relation to each other. No part is regarded solely as a preparation for the others.

b) The use of econometric models has been added.

c) The importance of the scenarios has been de-emphasized both as a tool of investigation and as a means of synthesizing various threads of information.

d) Work on producing a synthesized presentation has gained in importance (and has been performed outside the scenario framework).

Let us present the different parts I-IV one by one and then conclude with a discussion of the main changes made.

## I Background studies

If we compare the anticipated titles at the beginning with those finally published, we can see that the background studies have changed somewhat. A list of the <u>published</u> reports (in Swedish) is given in the table on the next page.

Publications

1. "Resources and Raw Materials - some starting points for a study" by the project group, 172 pp.
2. "On Natural Resources and Economy" by Carl-Axel Olsson, 64 pp.
3. "Remote Sensing and Natural Resources" by Claës Pilo, 86 pp.
4. "Renewable Raw Materials" by S. Sohlman, Erik Swedborg, Reidar Persson, Erich Jacoby, 176 pp.
5. "Aluminium - a Raw Materials Chain" by Sten Lindeberg and Uno Svedin, 141 pp.
6. "The Sea and its Resources" by Thomas Bertelman, Björn Ganning, Hans Ackefors, Bo Hall, Karin Hjertonsson, 162 pp.
7. "Politics and Raw Materials in an International Perspective" by Tomas Bertelman, 220 pp.
8. "Recycling of Raw Materials" ed. Staffan Sohlman, 78 pp.
9. "Substitution - to Exchange Resources" by Uno Svedin, Tommy Bengtsson, Göran Grimvall och Karl G. Jungenfeldt, 240 pp.
10. "Land - a Scarce Resource" by Tomas Lindencrona, 110 pp.
11. "Non-renewable Raw Materials" by Ernst Hollander, 256 pp.

In retrospect several observations could be made regarding the use of the background studies, quite apart from the overall observation that ultimately they certainly became one of the major occupations of the project group and provided the readers with a small library of more than 1500 pages devoted to raw materials issues. Some comments on developments during the lifetime of the project in this specific area of work are:

- The background studies were initially envisaged as being papers of a "state of the art" nature. However, in their published form they were more "complete" as they also include broad presentations of future prospects.
- The published "background studies" are also more "complete" in other respects i.e. in that they treat a broad spectrum of aspects. For example the "Resources of the sea" study includes aspects of mineral resources on the sea floor, living

resources in the water, environmental aspects and aspects of international law. Thus the theme anticipated from the beginning "The sea and sea-floor resources" has been extended to encompass environmental and juridical aspects as well.

- The full set of published background studies has a pattern of its own. In the beginning the overall pattern was not at all obvious. The non-existence of a real pattern reflected different views in the project group and the reference group about where efforts should be concentrated. In the reference group there was a wish to give the background material a more Sweden-oriented and survey-like character. The final structure is geared more to covering "overall problem types". For example there was a wish to have more raw material chains investigated. The project group developed one prototype (Aluminium). In the initially anticipated scheme one interim study was to cover "geopolitical descriptions of what is available". Instead of such a survey, an analysis was made of what is meant by "available resources", as well as a study of the estimated availability of individual mineral resources on a global level. A study was also made of the politics of raw materials.

- Most of the anticipated studies were in fact carried out in one way or another. Some were more complete than others, however, indicating that priorities changed during the life time of the project. For example:

  - "Analysis of Swedish raw material policy" resulted in two unpublished studies of some considerable length: "Policy of natural resources in Sweden 1945-1970" and "Raw materials in Swedish foreign trade". These reports were never published because it was felt that the subjects were too specialized and had perhaps been written in the kind of language that was not very accessible to a broader readership. This thus reflects the project groups reaction to the idea of the two categories of readers discussed above - the public in general and the decision makers. The substance of the studies, however, was included in the final report.

  - "Energy-need for raw material production" became two reports: "The supply of raw materials in a perspective of energy shortages" and "Energy - a useful concept in raw materials accounting". The theme "Energy and raw materials" obviously called for a combined effort between the "Energy project" (described in Section 4.1) and the "Resources/raw material project". The two projects represented different views as to the issues which most urgently needed clarification. The energy project group was concerned to find out whether the availability of crucial raw materials might somehow become a constraint on the energy production technologies of the future (e.g. the availability of stainless steel and other special substances of various kinds). They also wanted to know in energy-budget terms how much energy would be needed to produce the raw materials used in the energy production and distribution systems.

The raw materials group on the other hand wanted to know whether energy shortages might become a serious constraint on various raw materials uses.

The reports turned out to be fairly technical. Due to the obvious interest of a broader readership, publication through other channels was promoted. On the energy side it was the raw-material-accounting aspect that was thought to be important, and it was felt that this paper should provide the basis for part of a broader study of accounting issues under the heading "Resources as flows". Partly due to lack of time, partly due to the theme´s more conceptual character, this study was not finished. Parts of it have been kept available as background material.

- Environment. Here a number of subsidiary papers were written by various consultants, with a view to ultimate integration into a study of "Raw material and the environment" or something similar (the topic was not on the original list). Due to the diverse character of what was written, and the lack of time in view of the need to amalgamate all the parts, the study was not published. The main factor contributing to this failure was the illness and later the death of one of the core group members - the ecologist.

- Some <u>new</u> sub-items were considered in the course of the work on the project, some of which had been used, for example, to substantiate discussions or to serve as general information collectors. Examples are studies of Japanese raw materials policy, new raw material technologies, raw materials market structures, etc.

To summarize developments regarding the background studies we could say that:

| | | | | |
|---|---|---|---|---|
| - they turned out to be | less | data-oriented | and more | concept-oriented |
| - -"- | -"- | mapping -"- | -"- | problem -"- |
| - -"- | -"- | Sweden -"- | -"- | global -"- |
| - -"- | -"- | discussing present situation | -"- | directed towards (conventional) futures of various sectors. |

## II Scenario work

Turning now to the work on the scenarios, three phases could in retrospect be distinguished (see below). The team of authors was not happy even with the final result, despite the considerable effort they had put into it.

| Phase | Type of scenarios | Problems |
|---|---|---|
| 1. | General global political | Too little connection to raw materials. |
| 2. | Strongly and directly linked to raw materials (wood, hard metal etc) | Too weak a linkage with the general political framework. |
| 3. | An attempt to describe the surrounding world and at the same time separate developments for individual raw materials sectors. | Problems of consistency. |

In phase 3 three basic types of "surrounding worlds" were developed in the form of macro-scenarios:

- business as usual (as a reference-scenario)
- global north-south conflict
- global cooperation.

The choice of scenario alternatives presented quite a few problems, particularly the choice of the main parameter to be varied between the scenarios.

What is really meant by "business as usual"? Is it really possible to imagine that all the tendencies of the 1960s and 1970s will continue - and can continue (e.g. growth in raw materials consumption and in GNP, etc)? Isn´t such a picture of the future full of internal inconsistencies which in themselves make the scenario improbable? Or does "business as usual" mean that the structure can be modified without losing its decisive characteristics and thereby maintaining some similarity with today's world (e.g. the existence of a market economy, a certain type of international division of labour etc)? But if some modification is allowed, then the rate of change will be the important factor. How much can the structure be allowed to change? And what is the character of the pattern of forces which acts upon the structure? And why did just the chosen set of forces determine the outcome and not others?

A number of questions of this kind were raised. The need to base the discussions on some sort of theory of societal change became evident. However, many of the participants in the project group felt that existing models were in one way or another too narrow in scope. And in many cases the theories differed sharply due to differences in their underlying assumptions.

But already before the content of the three surrounding-world scenarios had been worked out, problems had arisen. Perhaps the international division of labour (the choice reflected in the three scenarios listed above) was not the most suitable main parameter. Perhaps other variables such as the rate of economic growth would have been a better choice? Possible candidates for scenarios within such a framework could have been:

- continued growth
- zero growth
- economic collapse.

Or perhaps yet other variables were particularly important, e.g. the availability of energy or the capacity of technology and social organization to cope with environmental degradation. Thus, already at the macro level problems appeared in the choice and design of "future surrounding worlds".

In the end the international division of labour was in fact selected as the scenario parameter. One important reason for this choice was the observation that in the case of food, and especially as bearing on the situation in the developing countries, production and distribution are intimately connected. Here we have a reflection of one of the basic value assumptions guiding the study, namely the interest in questions of global equity.

Once the choice of the main scenario parameter had been made and the different macro-scenarios outlined accordingly, it was time to let these scenarios colour the various descriptions of developments in connection with individual raw materials (which for conveniance were lumped together in groups such as crops, wood, metals used in stainless steel-alloys, hard metals etc (see below). This work involved describing the elements of the following matrix.

| | | Macroscenarios | | | |
|---|---|---|---|---|---|
| | | Continuation of present day trends | Cooperation internationally | | Conflict between developed and developing countries |
| | | | in a selective way between various countries and regions in developed and developing countries | through a supranational network with the goal of global development | |
| Raw-materials scenarios | Hard metal | | | | |
| | Stainless steel | | | | |
| | Aluminium and copper | | | | |
| | Forest products | | | | |
| | Iron | | | | |
| | Cereals | | | | |

At this stage a problem that arose was the lack of interesting outcomes of the exercise. In most cases the variations between the macro-scenarios did not make much difference to the outcome of the individual raw material scenarios. (In some cases, however, this could be regarded as an interesting result in itself).

A feeling arose in the project group that the results were arbitrary, and that in most cases alternative designs of the scenarios in the essay-like form that was used were perfectly possible. Thus, scenarios did not help to investigate future outcomes of raw-material situations in the world, at best they gave some idea about where the problems might lie. But this problem-identification job could probably have been done better in other ways.

Finally the group felt uneasy about using the scenarios as result-conveyors i.e. as a way of presenting results. This uneasiness was born of the fear that their own awareness of the arbitrariness of the "results" might not be obvious to the reader in its full extent. There would thus be a not inconsiderable risk of conveying a false impression of the validity of the arguments. In this context a method involving descriptive scenarios interwowen with textual commentaries was discussed but never tried.

## III Studies using econometric models

To try to tackle the problem of inconsistency in the use of essay-like scenarios, formal computerized econometric models were introduced.

The analysis was performed in several steps which included the use of two separate models developed in other contexts. The stages were:

1. Different global macro-scenarios were sketched, similar to those produced in the scenario work.

2. These scenarios were translated into input data for the econometric model developped for the Swedish long-term economic surveys (Restad, 1976).

3. Computer runs of the econometric model were performed. Results were generated in the form of alternative economic developments for Sweden.

4. The alternative economic developments were taken as input to a model which had been developed by a Swedish economic defence agency for analyses of the need for raw material stocks for war-time use.

5. The model mentioned under step 4 above generated sets of raw material needs, for each of the alternative general international developments outlined in point 1.

The method provides some insight into the differences in outcomes for raw materials due to variations between the original alternative world futures. These <u>differences</u> are less arbitrary and sensitive to limitations in the realism of the models than the <u>absolute levels</u> of outcome for individual raw materials. Here we find a certain formal rigour and consistency in the statements.

However, even these statements contain major elements of uncertainty, despite the deceptively precise character of the quantitative results. This is of course due to the implicit assumptions woven into the structure of the models. A long series of assumptions about economic and societal structure and changes may be hidden in the model structure. This could cause problems for the model user if he takes the resulting figures too literally.

Nor is it easy to introduce realistically changes in technological capabilities, or substitutions, into the structure of the model, and this reduces the model´s ability to provide reliable forecasts over long periods in which technological change can begin to make a fundamental structural impact.

The power of econometric models - or of any formalized models - lies in the control which they provide over the covariation of the factors involved. The weakness of the method is that certain elements, often of a non-quantifiable type and reflecting changes in structure, are seldom considered within the framework of the model. As indicated above, the dynamics of the real world cannot be entirely reflected within the apparatus of the models. Consequently forecasts based on these models have to be regarded with a certain amount of healthy scepticism. The models cannot do more than simply tell us what we could expect to happen, given a certain set of assumptions and assuming limited change in underlying structures.

## IV Work on a synthesis

The last quarter of the time spent on the project was entirely devoted to producing a synthesis of all that had been learnt hitherto. To give a very brief idea of what type of material was considered, we reproduce below the headings in this final report (Bertelman et al, 1977, 1980):

- introduction (in which the international debate was reflected as well as some more methodological considerations regarding the role of a futures study)
- what is meant by resources and raw materials?
- projections of global demand
- projections of global production
- the distant future
- Sweden and the future raw-materials situation
- goals and means for a rational raw-materials policy.

A few comments could be made on this synthesis.

On interdisciplinary work

First of all synthesis is more than simply adding pieces together. The whole is more than the sum of the parts. This point was especially important in the present case as one of the underlying aims of the study was that it should be of an interdisciplinary and cross-sectoral kind.

Only the members of the project group participated directly in writing the final report. However, the manuscript was also commented upon by a few external experts (including one whose job was to see that the language was kept simple).

It is often claimed that the lack of a common language restricts trans-disciplinary work. In fact, once the various technicalities have been explained, it is probably not the language itself that provides the stumbling block; much more important are differences in the direction of interest. Social scientists are more interested in the questions of distribution, of social structure - in short, the question "for whom?". The technicians want to tackle questions like "how is it done" or "how could it be done".

Theory basis

Working on a synthesis also reopens the question of the theoretical basis of the project. As we have discussed earlier, most members of the project group were inclined to be sceptical of too much theory. Theories could and should be used in parts where they appeared appropriate, and all the assumptions should then be made explicit, but theories were never to be allowed to provide the overriding framework for the whole study. Closely related to this issue is the question of the proper balance between "soft" and "hard" sources of information, and closely associated with this is also the conflict between those who prefer formal methods and those who like more "essay-like" approaches. A discussion in the project group on the usefulness of presenting a poem in the text illustrates the range of opinion on this point.

Explanation of change

Another problem concerned the attitude towards historical explanations of change. Was it the invention of the iron plough (i.e. a technical invention) that made deeper furrows possible and thus improved production possibilities? And was it the nature and shape of the new agricultureal plot (resulting from the technical constraints) that generated different frameworks for the economic and social structure?

Or was the reverse process true? Perhaps instead we should start with social and economic factors, and then see how they change the circumstances for technical development. Was it the social and economic situation that made it possible for the industrial revolution to start in just England and Germany? Or was it man´s attitude to the relation between himself and nature that made this development possible in Europe, but not in India? Questions of this "hen and the egg" type arise as soon as we start discussing major changes of global significance. Either we don´t answer the question at all and limit ourselves to describing the situation before and after. Or we name some external factors that we believe to have decisive implications. Or we believe that many changes, insignificant in themselves, can together constitute a significant change in the system as a whole. In any case, any project group will have to tackle these issues.

Overriding variables, Values

Other questions that had to be seriously discussed concerned the dominating variable in different time frameworks: was it economic necessity, social structure or psychological or ecological constraints? This issue brings us back to the discussion at the beginning of this chapter about the spectrum of values represented in the project group - many of which could not be put into words at the start.

Let us round off this description of the development of a futures studies project by summing up a few aspects (but important ones) in which some change was obviously taking place in the course of the work.

## Changes during the project

Many of the factors originally motivating the study reflected economic conditions (rising prices, limited stocks, cartels etc) and as mentioned above economic tools of analysis were used (together with other approaches). The economic focus had also a decisive impact on the structure of the final report, in which two central chapters are entitled "Global demand" and "Global supply". However, the discussion about values during phases 1 and 2, and especially the discussion of international solidarity, shifted the focus towards international economic and social development aspects to a greater extent than a simple and straightforward examination of Swedish economic futures would have required. Technical and ecological aspects were also given more space in the final report than was originally planned, as a result of discussions that took place at later stages in the study.

### Time span

In phase 1 nothing was really said about the time span. Perhaps, however, certain assumptions were implicit in some statements in the original Secretariat project document, e.g. that futures studies should provide a basis for choice and cover developments where present-day constraints have softened; that the rights of future generations should be observed. However, in the early stages the members of the project group thought this would mean

10 years, plus a complementary outlook up to a 25 years horizon. Later the span was to grow considerably. In the final report the main focus is on 15-20 years, plus a long-term view of perhaps 25-70 years. In fact in the final version the perspective was very much more flexible than had at first been envisaged.

Balance between a national and an international perspective

The original strict focus on Sweden gradually changed and descriptions of a more global nature became more important. Given that perspective, some conclusions affecting Sweden were then drawn. As we have seen, this probably went against the initial wishes of the reference group, which wanted a rather matter-of-fact (and shorter-term) study, strictly relevant to Sweden.

Integration of sectors, disciplines and scientific traditions

In the various reports and background studies the major sectors were kept separate (agriculture, fisheries, forestry, mining). This was also true of the final report. Thus in one sense integration has meant no more than collecting and juxtaposing the various presentations. On the other hand, this very compilation provided opportunities for cross-sectoral comparisons which have certain integrative effects. The same can be said of the general background discussions.

Various disciplines have made contributions to different parts of the text. The balance finally struck between economic aspects and others of a more ecological nature, for example, will have to be judged by the readers. However, there was an ambition in the group to discuss such problems without prejudice.

Balance between normative and explorative aspects

The normative element in the Secretariat´s project document is obvious ("opportunity to choose", "alternatives"). In the final report, however there is perhaps less stress on specific normatively chosen futures; instead there is a critical examination of various statements about such futures, providing the reader with an idea of the kind of problems involved in judging the future within this field.

No visionary alternatives are given for the far distant future. In fact there is a tendency in the text to advise against the use of visionary pictures without first securing a reasonably convincing path from the present to the particular future point in time under consideration. The tone of the initial task is matched less by normative descriptions than by choice-oriented analytical scrutiny of problems involved in the assessment. Efforts concerned with "choosing the future" can best be supported by providing a set of "possible tools" to use in developing a natural-resource and raw-materials policy. But the authors do not hesitate to support the initially included normative value assumption of international solidarity - a theme which in fact flavours the entire final report.

In the foreword to the final report the authors explain: "we had of course hoped to say more and to say it more exactly". That more has not been said is explained by the limited time available for the work of the project group. In retrospect, however, one wonders just how much "non-saying" was due to internal disagreements on points of balance etc. This is of some interest, as it throws some light on the question of value judgement. The value judgement explicitly stated in the study certainly gave some direction to it, but not of such a kind as to create internal turmoil in the project group. It is obvious that the heated debates were caused by other things, but what exactly these were is difficult to pinpoint. Below is a possible list:

- the possibility of drawing conclusions from existing social science theories (e.g. Marxist theories)

- the value of presenting visionary pictures (sometimes known by their opponents as kingdom-come pictures)
- the balance between what are, ultimately, the important factors (e.g. are they ideas, or are they of a naturalistic origin - economic-societal or ecological, perhaps)
- what should be regarded as acceptable pieces of information (e.g. a poem?). How scientific should one be, and what is meant by scientific in the context of the study
- what is the attitude towards the knowledge acquired with the help of various analytical tools (scenarios, econometric models).

Attempts were made to present these difficulties and "hidden" discussions as openly as possible, the idea being that they probably reveal more about the method used than the meagre amount of information that generally accompanies the official presentation of a project.

At the same time it should be noted that the change in opinions on various aspects of the study material in the course of the procject should not be regarded as a sign of unwanted effects. On the contrary it is crucial in any futures study project to allow for the internal development of goals and methods. Indeed, it could be argued that a change in perspective during the course of a project is in many cases an expression of a growing understanding of the real nature of the issues.

## 4.3 TRANSPORTATION AND TRANSPORT RESEARCH - FUTURE DIRECTIONS

### Purpose

The Transportation Study which we will refer to here as the Trp study, was initiated by a government research council, the Swedish Transport Research Delegation, or TRD. To explain the purpose of the Trp study more clearly we can first briefly describe TRD itself. It is intended both to commission R & D projects and to finance R & D initiated by researchers. More generally, it is responsible for planning, co-ordinating, initiating and funding R & D in the transportation field, in areas which are not the responsibility of any other government agencies. Other agencies with responsibility for commissioning research in the field of transportation are, for instance, the Road Transportation Agency and the National Railways. TRD's research responsibilities concern traffic safety, for instance, as well as problems which cut across different modes of travelling and problems connected with transportation policies on the national level. TRD had at this time a yearly budget of about 12 million Sw. Crowns (about 2,5 million dollars).

TRD was established in 1971. During its first years most of its budget was directed towards rather short-term operational R & D problems. After a while it was felt that more resources should be spent on longer-term problems, and this seems to have been a main reason why TRD initiated the Trp study in 1975. In an early TRD document, written before the project group was recruited, the general scope of the project was outlined. Several purposes for the study were stated. For instance, it should indicate areas where major changes could affect developments in the transportation sector in Sweden. And, most importantly, it should provide a basis for TRD's research policy and help to indicate directions for long-term R & D work in the field of transportation.

In the TRD project document the Trp study was referred to as a "prospective plan". However, it was pointed out that no detailed prospective planning model existed, ready to be applied. In principle the study was to cover such aspects as the formulation of goals, problem analysis, forecasting, alternative futures for different sectors of society, future transport needs, and the development of transport technology. But no mention was made of how it would be possible to identify R & D needs on a basis of this type of planning study.

Prospective planning is not necessarily useful in pointing out directions for research. In general, it seems reasonable to expect that future-oriented studies should be useful in anticipating problems. If problems are foreseen, we have time to identify gaps in our knowledge and to carry out such R & D as can help us to deal adequately with the problems. But why should a future-oriented study take the form of a prospective plan? On this point it seems that some of the ideas behind the initiation of the Trp study can be traced back to a committee report "Research for the Defence Sector" (SOU 1970:54), which stresses the need for a close link between long-range defence planning, so-called prospective planning, and R & D. In the Swedish defence sector, prospective planning involves the analysis of future defence goals and defence postures and the outlining of alternative strategies of development. It can thus be seen as an equivalent to strategic planning in private enterprises (Ansoff, 1965; Anthony, 1965). Its purpose is to provide a basis for the decisions of the Riksdag, which come up every 4 or 5 years, on the long-term development of defence. As the prospective plan for defence is not a plan to be implemented but merely a general framework within which future developments are considered, it can be regarded as a kind of futures study if we apply Shani's classification schema for distinguishing between futures studies and conventional planning (Shani, 1974).

To summarize: a main purpose of the Trp study, as formulated before the project started, was to provide a basis for TRD's research policy and to help to point the way for its R & D efforts. A basic, but actually quite uncertain hypothesis behind the study was that some kind of futures study or prospective planning of the transportation sector could provide a better overview of the need for R & D.

### Project organizaion

The Trp study was a much smaller project than the Energy and Society and the Resources and Raw Materials studies. The project group consisted of two people, working less than full time for 2 1/2 years. There was also a reference group for the project. This reference group, appointed by TRD, had six members, three of whom were also members of TRD. The project group and the reference group met for discussions a couple of times each year. The reference group possessed considerable professional knowledge in the field of transport research and transport policy. It included several professors with long experience of transport research, an administrator from the Ministry of Transport, etc. The project group had no previous experience of transport research but were specialists in long-range planning methodology and systems analysis. They also had general experience of various areas relevant to transport research (economics, statistics, technical areas etc). They participated in the project as part of a broader research programme concerned with the methodology of planning and futures studies.

### General approach

As the project members were experienced in planning theory, they found it natural to choose a planning approach and to start by examining existing conceptual models of prospective or strategic planning. But it seemed to them questionable how far existing models were relevant to the transportation sector and to the particular purpose of the Trp study. Because of considerable uncertainty in these respects the group suggested dividing the project into two sequential steps. The first phase would be a preliminary or pilot study, covering the whole problem area very roughly. The experience gleaned from this preliminary study could then be used to revise the strategic planning model employed during phase 1. Something could also be learned about what background material was available and about the amount of detail needed in various parts of the study.

This general approach, together with the purpose of the study, was discussed with the reference group a few weeks after the start of the project. There was general agreement on the cautious two-step approach suggested. As to the purpose of the study, the reference group insisted that the study should result in an easily readable product which could reach wide target groups, not only transport researchers and planners but also participants in the transport policy debate. It was questioned whether this dual purpose - to provide a future-oriented background for the transport policy debate and to help to indicate directions for R & D - might not impose different and conflicting requirements on the general outline of the study. At the time it was thought that this would probably not be so, at least as far as the first phase of the study was concerned. Later it became evident that the different purposes did impose conflicting requirements. As we will discuss later, the problems that arose were probably symptoms of differing views as to the priorities between the two purposes.

### Phase 1

A general hypothesis behind the chosen planning approach was that it would be both feasible and useful to distinguish between the *planning object*, i.e. the tranportation sector, referred to as the transport system, and the environment of the planning object, i.e. such aspects of society as affect the future development of transport demand and transport requirements. A number af such aspects were identified, for instance population changes, employment, working hours, urban structure, regional development, structural and volume changes in different fields of production, exports and imports etc. The conceptual model, that was applied in the preliminary study, is outlined in Figure 4:2.

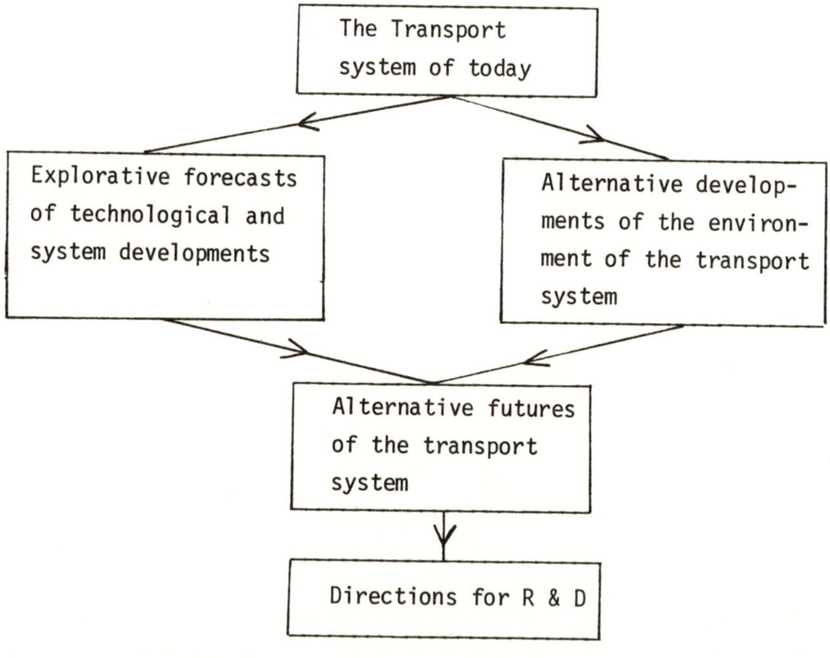


Figure 4:2 The phase 1 planning model

A number of problems surfaced during phase 1, leading to several changes in this conceptual model at the start of phase 2. Some of the main issues discussed during phase 1 were:

- Can a normative planning approach be used?

- What background material is available? Need for special background studies?
- Feasibility of forecasting future transport flows?
- How can research needs be deduced from studies of alternative developments of the transport system?

A normative approach?

It was obvious from the original mission description that some kind of normative study was wanted, i.e. the task was not to predict the development of the transport system but to consider desirable developments. But what is desirable, and for whom?

While there seemed to be general agreement on the use of a normative approach, there were also discernible differences of opinion as to the "level" at which this approach should be applied. Was it feasible to design desirable futures for Swedish society and future alternatives of the transport system that fitted such societal futures? Or should a number of desirable developments or directions be designed for the transport system? Or should norms and standards be formulated for various parts of the transport system concerning, for instance, accessibility to public transport, safety standard etc? The project group did not see the formulation of specified norms and standards as suitable tasks for research; these were political issues. Nevertheless, transport research literature often presupposes, explicitly or implicitly, some kind of transport goals. What were these goals, and could they be used in the Trp study?

In transport economics the focus is often on economic efficiency or on satisfying the demand for transport. Geographers usually have a somewhat different perspective. The time and spatial dimensions of different kinds of transport are emphasized. Satisfying the need, rather than the demand, for transport is often regarded as a goal. However, future need or demand for transport will depend on a variety of changes in society, affecting the non-transport as well as the transport sectors. For various reasons the satisfaction of the need or demand for transport was considered inadequate as a goal formulation. Goal formulations of this kind do not convey any concrete impression about the directions to be sought for future developments. Nor do they take account of the negative side-effects of transport, for instance road accidents and pollution. Future demands and future needs for transport depend on the future structure of production, services etc which cannot easily be forecast and which are dependent on various societal goals. So the goals of transportation policy need to be related to the goals of society.

The kind of normative approach that was eventually chosen in the Trp study was based on an examination of the various effects and side-effects of transport on society, and on an analysis of such goal conceptions in the areas concerned as could be deduced from recent government decisions and government guidelines for committee work etc. As a result, five goals or goal dimensions were identified:

- acceptable daily programmes
- better balanced regional development
- efficiency of production processes
- limitation of negative side-effects of transportation
- limitation of energy use.

By "acceptable daily programmes" was meant that the available transport facilities should allow every individual access from his or her residence to places of work, shopping facilities and public services with a reasonable choice of action alternatives.

We will comment below, when we outline the main study, on the use in the study of the identified goal dimensions. Suffice it to say in the present context that the examination of the normative issue had revealed considerable differences in outlook between different disciplines, differences which seemed to need some bridging if research results were to be readily applicable to real-world problems.

Commissioning of background studies?

Originally there were plans to commission quite a number of background studies from various researchers or experts. Obviously, the study would require expert knowledge in a number of quite different areas and a small project group could not be expected to cover all of them. However, during the first phase of the work it proved almost impossible to delimit problem areas in such a way as was needed to commission special studies. The reason was that a considerable number of studies such as research reports, committee reports, forecasts, plans etc, that contained material relevant to the Trp study, were already in existence. This information needed analysing and structuring in new ways, but exactly how depended on the logical structure of the Trp study itself, and this was still an open question. Later, it was easier to delimit certain problem areas; but at that stage it seemed doubtful whether it was worth trying to get access to experts for relatively small projects which would have had to be carried out at short notice. In the end only one quite small study was commissioned. It would probably have been an advantage if the study could have been organized so as to make access to experts easier.

Forecasting future transport flows?

It seemed to be a generally held view that a study of the Trp type should result in forecasts of future transport flows. These would include the flows of passengers and various types of goods, for different routes and different modes of travel. But can such forecasts be made in a long-term perspective, say up to the year 2000? Are such forecasts needed? How could they help to indicate directions for transportation research and development?

As to the usefulness of long-term quantitative forecasts, it cannot be denied that these are sometimes needed. For long-term investments, in new freeways or railways for instance, forecasts of future transport flows can be useful to those who have to decide about capacity and priorities. But capacity problems are not the only problems in the transport sector. And the necessary forecasts can often be made more easily in the context for which they are needed, so that the presuppositions can be fitted to the actual decision situations.

As to the Trp study, it seemed more important to try to identify problems than to estimate future quantities. Future transport flows will depend on a variety of factors such as transport policy, taxes, regional policy, changing industrial or production structure, working hours etc. Obviously, no very accurate long-term forecasts can be made. For the purpose of the Trp study it seemed more important to analyse the way in which various factors would influence future transport flows and to relate this to possible problems of development. In fact several quite comprehensive forecasts had been made on the initiative of the Ministry of Transport prior to the start of the Trp project. In the Trp study it was therefore possible to focus on an analysis of the assumptions underlying these forecasts and on the possible influence on the results of various assumptions. Among other things, factors important to the choice of modes of travel and possible changes in such factors were examined.

Included in the phase 1 model was a section devoted to "Alternative futures of the transport system". The original ideas of the project group about this part of the study did not directly concern future transport flows. Instead it was thought that this section would involve the design of different futures for the physical transport system and its use. This would be consonant with the concept of planning as defined by R. Ackoff:

> "Planning is the design of a desired future and of effective ways of bringing it about." (Ackoff, 1970)

The importance of designing alternative futures for systems or planning objects that include components which make changes difficult, has often been discussed (Schwarz, 1976). By designing different futures, which may be desirable under various set of assumptions about future developments, it is thought that it will be easier to anticipate the necessary changes in time.

But would it really be wortwhile to design a number of futures for the transport system? This was discussed in an internal memorandum some six months after the start of the project. A great deal of detail in the descriptions of the futures would be needed for evaluating them. In view of this amount of detail, the number of different futures needed to cover the different possibilities seemed endless. And, in the case of the transport system, do we really need to outline several futures at a distant planning horizon? This planning idea has been developed for systems which can only be changed slowly. But in the field of transportation some quite big changes can be made overnight: the same cars can be used to transport different goods on different roads, buses and trains can be used for different transport etc. True, other components such as roads and railways take more time to change, but they are not so closely connected with all the other components of the transport system as to make it essential to design different futures for the entire transport system.

At the end of phase 1 it was decided that the Trp study should not focus on quantitative forecasts of future transport flows, nor should alternative futures be designed for the entire transport system. Instead, a new element, "plans and trends", was added which included among other things an analysis of the principles for road investment.

How can transport research needs be deduced?

A crucial issue was of course whether long-term R & D needs could actually be identified with the help of prospective planning or a futures study of the development of the transport system. As we have mentioned before, this was the basic hypothesis underlying the Trp study. In fact the need for linking sectoral R & D policies to sectoral planning and social goals had been a central issue in the international science policy debate during the early 1970s.[1] But what form should this linkage take? Here there seems to be a clear difference between product-oriented and problem-oriented R & D. In technology-intensive areas, such as defence, in which long-range plans are used, it is reasonable to expect to be able to deduce from the planning process certain guidelines for R & D needs, particularly as regards the development part of R & D. In other areas where R & D requirements stem from problems rather than from technology-based systems, it is by no means obvious how problems can be anticipated with the help of a planning process or how planning can help to indicate directions for R & D. It was concluded from phase 1 of the Trp study that there was no way in which all the R & D needs of the transport sector could be deduced from a study of the Trp type. But as a result of the reorientation of the study after phase 1, the project group thought that the study might be able to provide an

1) See for instance the Brooks' report (OECD, 1971), "Science and Technology in the management of complex problems"(OECD, 1976) and "Linking Research Planning to Sectoral Planning" (Schwarz, 1979).

overview of a number of developments and problems connected with transportation, and also generate some guidelines for R & D efforts. It was also concluded that to serve this purpose the Trp study would have to be more explicitly problem-orientated.

Phase 2

After phase 1 the planning model was revised, see figure 4:3. It was stressed that the model should be considered primarily as a learning model, a logic for collecting and structuring information so as to obtain the perspective and knowledge needed for an overview of problems and R & D needs. The revision of the planning model was accepted by the reference group without any visible opposition, probably mainly because the suggestions had been introduced gradually. Priority had been given to the R & D-orientating purpose of the study. The other purpose, to provide a future-orientated background for the transport policy debate had been de-emphasized but not abandoned. It was still intended to present the material in a form that would be readable as well as interesting to a wide audience. Nevertheless, the possibility cannot be ruled out that the apparent consensus was partly due to the fact that some members of the reference group had stopped coming to the meetings. Perhaps the main interest of these members concerned the transport policy issues. And later, when the manuscript had already assumed its final form, considerable confusion arose about the mode of publication and the foreword.

The seven different parts of the phase 2 planning model correspond to separate chapters in the final report (Schwarz & Svensson, 1978). If we compare with the phase 1 model, we will find that one element, "Alternative futures of the transport system", has disappeared and some new elements, namely 1, 5 and 6, have been added.

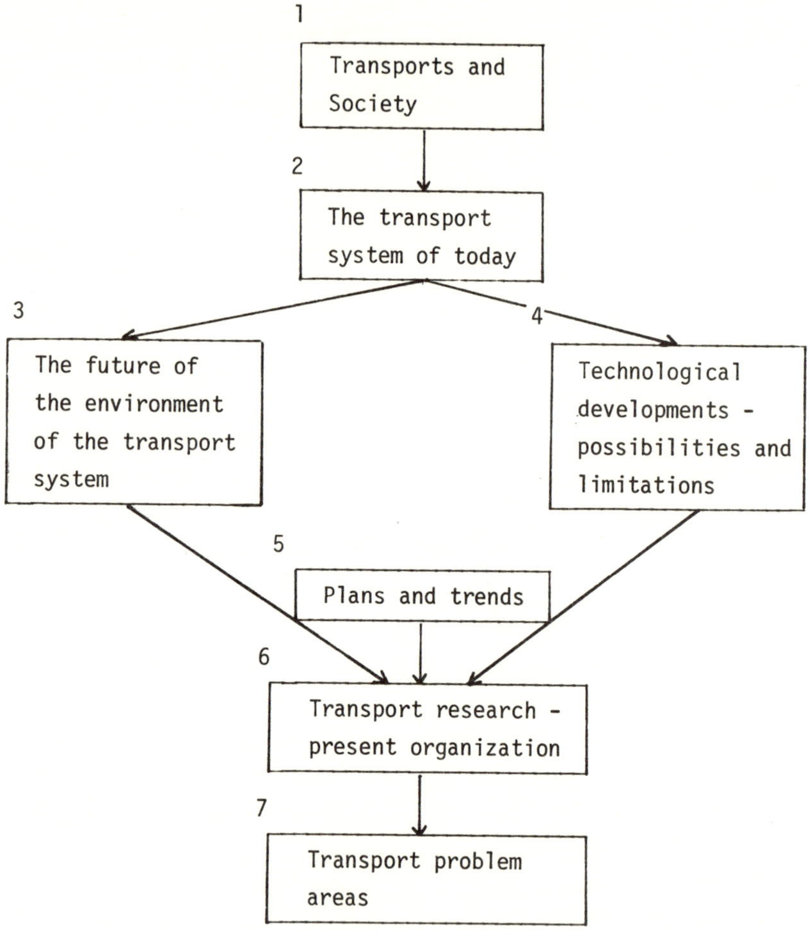


Figure 4:3. The phase 2 planning model.

Part 1, one of the new elements, consists of an analysis of the transport system as a system. The system includes both goods and passenger traffic and individual as well as public transport systems. The problem of formulating goals for the transport system is examined and, as was mentioned above in the discussion of the normative approach, five goals are defined.

In Part 2 the functioning of the present-day transport system in Sweden is described and analysed; the five goals mentioned earlier provide the structure of the analysis. Insufficient goal-fulfilment is considered to be the main problem of today's transport system. One of the difficulties is that the transport system tends to aggravate rather than to counteract existing regional imbalances. However, perhaps the most serious problem in the field of transportation today is the negative side-effects of transport, particularly road accidents and pollution.

In the Trp study the term <u>problem</u> refers to real-world problems, as in the examples mentioned above. Real-world problems are not the same as research problems. But if the real-world problems are thought to be serious, then it is worth examining them further to see whether more knowledge is needed to handle them, and whether R & D can contribute.

Future developments of importance to the transportation sector are examined in Parts 3 and 4. As mentioned earlier, the environment of the transport system consists of such aspects of society as are important to the development of the transport system. The idea was to outline in part 3 a set of scenarios describing alternative future developments of the environment of the transport system. However, in this context, a number of methodological problems had to be settled. Should quantitative, alternative forecasts be made of each factor, such as production in various branches and regions, weekly working hours, etc? Is there a relationship between the development of different factors? If so, how can different alternatives be combined in a consistent way? Can alternatives for all factors be combined into consistent scenarios, i.e. different future developments of the total environment of the transport system? If different scenarios are constructed, what factor(s) should vary between them? And how should the transport system be developed so as to fit a given scenario?

As to the different environmental factors, it was felt that studies of each of these separately could be used directly. Studies of possible future developments of each factor, sometimes including the formulation of quantitatively described alternatives, conveyed a general appreciation of future directions and consequences for the transport system. As to the relationship between the different environmental factors, this seemed in most cases to be fairly weak. This made it less important to combine the alternatives for the different factors into "consistent" scenarios.

As regards consistency, however, there was one obvious problem connected with various general economic developments. Most earlier forecasts of transport flows were based on, or heavily influenced by, trends from the 1960s. They thus assumed a higher future growth of production than was considered likely in the middle of the 1970s. Yet figures from such forecasts were used in studies which assumed a rapid decrease in weekly working hours (30 hours a week), less favourable terms-of-trade etc - in other words changes which would make the implicitly assumed growth in GNP less likely. Because of the risk of this type of inconsistency, it seemed important to examine reasonable combinations of alternatives for factors that could have an impact on general economic developments and on the increase in traffic. In light of these considerations three scenarios were constructed, each representing "a future Sweden" or, more accurately, an example of a development of (parts of) the environment of the Swedish transport system.

The characteristic features of the three scenarios are presented in Table 1. Scenario A is a kind of reference case, outlining a fairly "surprise-free" future. In several respects it follows earlier trends fairly closely. The reduction in weekly working hours is no faster than during the 1950s and the 1960s, resulting in 7 or 6.5 working hours a day (5 days a week) towards the end of the century. Scenario B includes a more marked reduction in weekly working hours, resulting in a slower rate of economic growth. Scenario C is characterized by energy-supply problems, increasing energy prices and changing terms-of-trade.

Table 1: The Scenarios

| | A | B | C |
|---|---|---|---|
| | Reference-case | Short working-week | Energy problems |
| Rapid reduction in weekly working hours | | x | |
| Energy supply problems | | | x |
| Changing terms-of-trade (- 2 % per year) | | | x |
| More balanced regional development | x | | |

Results from long-range econometric studies based on computer models provided "consistent" estimates about the general economic implications of the scenario assumptions. There is a system of long-term economic planning in operation in Sweden: every five years alternatives are studied for the development of the national economy, generally for a five-year period. In 1975, however, an additional study of development alternatives for the next 25 years was included (Restad, 1976). Econometric models are not generally thought to be very good instruments for long-term forecasting, because of the difficulty of allowing for structural and technological change. However, in this particular case a number of variables were determined outside the model, which allowed for greater flexibility. Moreover, what is interesting about the results is often the difference obtained when certain parameters are varied: such differences may elucidate important effects even when the absolute values may be misleading.

Results obtained from the long-term economic survey were used in the Trp study to illustrate Sweden's general economic development in the three scenarios. See Table 2. Some adjustments were made to take account of some differences in assumptions between the economic survey and the Trp scenarios.

Scenario A corresponds to an average yearly growth in GNP of about 3-4 %. A consequence of the bigger reduction in weekly working hours in scenario B is that both production and consumption grow more slowly. In scenario C the GNP may grow at about the same rate as in scenario A, but changes in the terms-of-trade and energy problems result in slow (or zero) growth in consumption.

Table 2: Development of production and consumption in the scenarios (quantitative examples, yearly percentage).

| | A(%) | B(%) | C(%) |
|---|---|---|---|
| GNP | 3,6 | 2,3 | 3,4 |
| Consumption | 3,0 | 1,8 | 2,0 |
| Private consumption | 3,0 | 1,5 | 1,0 |

It should be noted that the figures in Table 2 are intended as quantitative examples of economic developments in the three scenarios and not as definitions of the scenarios.

Part 4 of the Trp study examines technological developments that may be important to the transport system. An example is the possibility that new types of engines and fuels may be able to reduce pollution and lead to lower energy consumption. The presentation of development possibilities and of factors which may influence the technological development, can be seen as an instance of explorative technological forecasting.

In Part 5 - one of the elements of the study which were added in phase 2 - existing trends and plans for transport by road, rail, sea and air are examined. The reason for including this analysis was that, after looking at today's transport problems in Part 2 and examining changes which may affect the nature of these problems and possible ways of handling them in Parts 3 or 4, it became necessary to examine the extent to which various transport problems were already being dealt with by the transport agencies as part of their planning activities, before any remaining problems that might need special R & D efforts could be identified. Part 5 looks at the way long-term investments are handled by the transport agencies, examines trends in car ownership, etc. One of the conclusions is that although a very advanced planning system has been developed for road investments, there is a risk that effects which are difficult to quantify or to measure in monetary terms do not receive enough attention. Examples of such effects are negative side-effects (accidents, pollution, noise) and regional development effects.

Part 6 discusses transport research as an example of sectoral research, i.e. mission-oriented research for a government sector. The organization of transport research in Sweden and the role of the Transport Research Delegation (TFD) are described.

Part 7 deals with the research planning process and ways of categorizing R & D projects to facilitate the dialogue between researchers and the users of research results. It can be useful here to structure the R & D requirements along several dimensions, for instance research purpose, problem background, research users, and research area. In Part 7 problem area is taken as a basis for categorization. The above-mentioned goals are discussed and problems that hamper their fulfilment are identified. The choice of policy instruments for reducing energy consumption can be mentioned as an example of one such problem. We need to know more about the design and effect of different policy instruments. For instance, should special agreements be made with the automobile industry, should new rules be introduced to govern the import of motor vehicles, should the tax structure be differentiated so that motor vehicles with high fuel consumption are taxed more severely, etc?

Another type of problem area is related to the organization of the transportation sector and the principles of pricing and financing transport services. A number of problems in this area are discussed in Part 7.

Report and implementations

As mentioned in the introduction, a main purpose of the study was to provide a basis for TRD's research policy and to help to orientate long-term R & D work in the field of transportation. It was also intended to provide useful background material for transportation planning and for the general transport policy debate. With these goals in mind, the report was prepared for publication in book form and special efforts were made to make it accessible in style to different categories of readers. But at this stage some confusion arose about the mode of publication. It turned out that some administrators of the Transport Research Delegation had been expecting a publication in their recently started internal series of research reports and that they thus probably had been assuming a somewhat different readership from that envisaged by certain members of the reference group. The book was published after a delay of a couple of months.

Did the Trp study fulfil its purposes? This is of course difficult to judge at the present stage. Generally speaking, futures studies are difficult to evaluate. They are often supposed to influence decision-making in rather an indirect way. Remarks such as the following are frequently to be heard in discussions of their role in decision-making:

- Their role is to sensitize decision-makers to the long-term consequences of their decisions.

- Their role is to provide a framework for judging long-term consequences or to cast some light on future threats and possibilities.

In this particular case the primary task - to indicate directions for R & D - was fairly well defined. The general approach was adapted to fit this task. In order to comment on the feasibility of making a study of the Trp type with a view to orientating R & D, we can make two comparisons. The first of these is with a recent American futures study in the field of transportation. The American study appears to contain very little problem-orientated or R & D information - nor was this its purpose. A tentative conclusion therefore is that if futures studies or long-range sectoral planning are to help to improve R & D planning, very specific requirements must be made on the design of the planning process. The second comparison is with prospective planning in defence. As was mentioned earlier, the idea of trying prospective planning for the orientation of R & D in the transport sector seems to have originated from a report from "Research for the Defence Sector". In the Swedish defence sector efforts are made to let information from the prospective plans influence the planning of R & D. However, most defence research is carried out at a defence research institute by teams of permanently employed researchers. This has made it possible to formalize a dialogue between the military planners and the research planners, and to introduce a system of long-range research plans. The situation is thus different from that obtaining in the transportation sector. Most of the researchers working for TRD are doing so only temporarily and there is thus not the same opportunity for developing R & D plans as there is in defence. It can perhaps be concluded that even if the Trp study has generated results important to the orientation of R & D, implementation is obviously more difficult than in the defence sector. In the analysis of the influence of futures studies on decision-making, we will return to a discussion of the impacts of the Trp study (Section 5.2).

REFERENCES

ACKOFF, R., A Concept of Corporate Planning (John Wiley, New York, 1970).

ANSOFF, I., Corporate Strategy (John Wiley, New York, 1965)

ANTHONY, R., Planning and Control Systems - A Framework for Analysis (Harvard University, Boston, 1965).

BERTELMAN, T., E. Hollander, C-A Olsson, M. Parmsund, S. Sohlman, U. Svedin, Resources, Society, and the Future, Secretariat for Futures Studies (Liber, Stockholm 1977, and Pergamon Press 1980).

JOHANSSON, T.B. and P. STEEN, Solar Sweden, Secretariat for Futures Studies (Stockholm, 1978).

LÖNNROTH, M., T.B. JOHANSSON and P. STEEN, Solar Versus Nuclear: Choosing Energy Futures (Pergamon Press, Oxford, 1980a)

LÖNNROTH, M., T.B. JOHANSSON and P. STEEN, Energy in Transition: A Report on Energy Policy and Future Options, (University of California Press, Berkeley, Los Angeles and London, 1980b)

NORDHAUS, W.D, "The Allocation of Energy Resources", *Brookings Papers on Economic Activity*, Vol. 4 (1973) pp. 529-576.

OECD, Science, Growth and Society, Report of the Brooks' Committee (Paris, 1971).

OECD, Science and Technology in the Management of Complex Problems (Paris, 1976).

RESTAD, T., Modeller för samhällsekonomisk perspektivplanering, SOU 1976:51 (Ministry of Finance, Stockholm, 1976).

SCHWARZ, B. & J-E SVENSSON, Transportation and Transport Research - Future Directions (In Swedish, "Transporter och transportforskning - Ett framtidsperspektiv", SP 1978:5, EFI, Stockholm).

SCHWARZ, B., "Some Conceptual Problems in Long-Range Planning" in Knowledge and Concepts in Futures Studies, S. Schwarz (ed), (Westview Press, Boulder Colorado, 1976)

SCHWARZ, B., "Linking Research Planning to Sectoral Planning" in Perspectives on Research, Development and Technological Innovation, D. Sahal (ed) (Lexington Book Program, D.C. Heath & Co, Massachusetts, 1979).

SHANI, M.,"Futures Studies Versus Planning", *OMEGA*, Vol. 2, No. 5, 1974.

SIMMONDS, W.H.C., "The Nature of Futures Problems" pp. 13-26 in H.A. Linstone and W.H.C. Simmonds (eds) Futures Research: New Direction. (Addison Wesley, Reading, Massachusetts, 1977).

SOU 1970:54, Research for the Defence Sector (Ministry of Defence, Stockholm, 1970).

SOU 1975:89, Långtidsutredningen 1975 (Ministry of Finance, Stockholm, 1975).

WITTROCK, B., "Long-range Forecasting and Policy-Making" pp. 267-288 in T. Whiston (ed) The Uses and Abuses of Forecasting (Macmillan, London and Holmes and Meier, New York, 1979).

WITTROCK, B., "Futures Studies Without a Planning Subject" pp. 119-150 in P.R. Baehr and B. Wittrock (eds) (Policy Analysis and Policy Innovation (Sage Publications, London and Beverly Hills, 1981).

A Time to Choose (Ballinger, Cambridge, Massachusetts, 1974).

# 5
# Problems and Experiences

The evolution of futures studies has not been without its controversies. Criticism of methods and practices is not of course unusual in connection with the development of a new field of intellectual endeavour. However, in the case of futures studies even the apparently basic assumptions and premises have been questioned. Is the future knowable? Can it be predicted?

The primary purpose of this chapter is to analyse and compare the three projects described in Chapter 4, to see what can be learnt from them on the methodological front. As was mentioned in Chapter 3, we use the term method as a generic concept encompassing not only techniques of analysis but also problems pertaining to the organization of knowledge and competence in the course of the study and in the presentation of the final results. However, to structure our discussion we have chosen a number of general issues as our point of departure. We have formulated these as follows:

- Is the role of futures studies to predict the future?
- Can and should futures studies contribute an improvement in decision-making?
- Problems of problem formulation, access to knowledge, and synthesis.
- Scenarios and models - choices and criteria.
- Values and the study of the future.

We shall address each of these issues separately in the following sections. However, it may seem premature to embark on a discussion of specific issues within the field of futures studies, if the basic prerequisites for the existence or acceptability of the field do not hold. Is the future knowable? And if it is not, does this mean that futures studies are futile? Or can such studies legitimately claim to be the product of something that is at least more than sheer guess-work?

The idea that the future is unknowable is usually connected with the view that knowledge must consist of true and well-founded statements about facts, and that facts are phenomena that can be observed; as future events cannot be observed, we cannot have any knowledge of the future. But when futurists argue that the future is to some extent knowable, they are either defining facts differently or using a different concept of knowledge (or both). For instance, Edward Cornish, president of the World Future Society, claims that some statements about the future may be at least as well-founded and certain, if not more so, than a statement such as "Christopher Columbus discovered America in 1492" (Cornish, 1977, p. 95).

The accuracy of statements about the future will naturally depend very much on the particular aspects of the future, or on whose future, is being considered. On a basis of theories of celestial mechanics, we can make predictions about the future movements of the planets which are likely to be quite accurate. When it comes to the future of complex social systems, our "knowledge" of the future, even the close-term future, is usually vaguer and much more conditional; but this is also often true of our knowledge of the present or the past in the same areas.

The possibility that forecasts may be self-defeating or self-fulfilling means that in many cases we can have no absolute knowledge of the future in the sense of being able to predict it. This is sometimes used as an argument against making futures studies. However, the reason why forecasts may be self-defeating or self-fulfilling is that we can influence the future by our decisions. And the idea that the future is not predetermined but can be changed by our decisions has in fact been one of the main arguments for carrying out futures studies. For instance, in a Prospectus for an Institute for the Future in the U.S. (1966) we can read:

> "The fatalistic view that the future is unforeseable and inevitable is being abandoned. It is being recognized that there are a multitude of possible futures and that appropriate intervention can make a difference in their probabilities. This raises the exploration of the future, and the search for ways to influence its direction, to activities of great social responsibility."

As a basis for the following discussion of our three case studies, we accept here the assumption that something can and should be "known" about the future. At the same time we will try to relate our discussion to a number of general issues in futures studies.

## 5.1 IS THE ROLE OF FUTURES STUDIES TO PREDICT?

All our three case studies - Energy and Society, Resources and Raw Materials, and Transportation - deal with past and present developments as well as those of the future. However, in none of the cases do the future-orientated parts of the studies result in a presentation of results labelled "forecasts" or "predictions". In fact the term prediction does not seem to be used at all in the studies, and the term forecast only when forecasts made by others are being examined. This brings up certain questions about the relationship between futures studies, forecasting and prediction, and about the role of futures studies.

The reader knows from our description in Chapter 2 that we regard futures studies as a broad field of future-orientated studies and not just as a kind of forecasting. However, there seems to be no general consensus about the delimitation of the field of futures studies, nor about the definition of such concepts as forecasting and prediction. But prediction is often taken to mean a fairly uncontroversial statement, intended to be accurate, about what will happen in the future. It is the usual term when the statement is based on what is regarded as a validated scientific theory. Forecasting often refers to fairly conditional statements about future developments, usually probable developments. This would be in line with the use of the term by such well-known "forecasters" as Martino (1973) and Ayres (1979). Tha latter states

> "Regrettably the discipline of forecasting is widely misunderstood, not only by the public but by many practitioners. The idea still lurks, both in our culture and in other cultures, that the future is preordained - it exists, though hidden from ordinary mortals".
>
> "Since the future is conceived as indeterminate to some degree, a forecaster deals with possibilities and probabilities, not certainties". (Ayres, 1979, Preface).

Martino (1973) has emphasized that the criterion of a good forecast should be its usefulness for decision-making and not its accuracy, i.e. not necessarily its ability to predict.

If forecasting is envisaged as dealing mainly with possibilities and probabilities rather than predictions based on deterministic theories, then the fields of futures studies and forecasting clearly overlap to some extent. Bertrand de Jouvenel, the well-known French futurist, sees the task of futures studies as being to clarify the range of possible futures and to create images of attainable and desirable futures. An American futurist, John McHale, defines futures studies as follows:

> "Futures studies is used to convey the idea of an activity which embraces many elements: prediction, conjecture, imaginative extrapolation, normative projection" (McHale, 1976).

Dror has pointed out some basic assumptions that are important if futures studies are to be accepted as a valid professional and scientific endeavour (Dror, 1975a). One of them is that something can be known about the future; he refers to the existence of "some stability or ultrastability of relevant phenomena" and argues that "without some regularity of phenomena no nomographic rules can exist, and such rules, in various ways, constitute the base for predictions". However, even if continutity in the development of many aspects of society is necessary to prediction and to the feasibility of futures studies, it does not necessarily follow that a futures study project is solely or even mainly aimed at prediction. The importance of other roles for futures studies has been stressed by Roy Amara, the president of the Institute for the Future, in a discussion of the objectives of futures studies. According to Amara six particularly noteworthy objectives of futures studies can be distinguished: "The first is to identify and examine alternative futures". ... "The second objective is to characterize the degree of uncertainty associated with each possibility or alternative future. ... The third objective is to identify key areas which are precursors or warnings of particular futures. ... Fourth we want to examine a variety of if ... then sequences. ... A fifth objective is to acquire an understanding of the underlying processes of change. ... Finally, the sixth objective is to sharpen our knowledge and understanding of our preferences." (Amara, 1978, p. 42).

The future events and developments described in a futures study may thus be of many different kinds. We will here distinguish in particular between the following:

a) developments that are probable under certain conditions (forecasts)

b) desirable developments (normative projections)

c) boundary developments (delimiting a space of probable or possible developments)

d) reference cases, e.g. obtained with the help of trend extrapolations

e) chains of events, e.g. designed in order to illustrate possible causal relationships.

As futures studies may thus involve different types of futures, it is naturally important to make it quite clear to the reader what type of future is being dealt with. It would obviously be misleading if a study dealing with probable futures is understood by the reader as dealing with desirable futures, or vice versa.

We have found a number of similarities in our three case studies, e.g.

- the result of the studies are not labelled as forecasts or predictions, although they include elements of type (a) above which could be called forecasts
- the studies all include most of the five types of futures outlined in a-e above

But the different types of futures have different roles and assume different forms in the three studies. Some examples from each of the studies follow.

## The Energy and Society Study

"To acquire an understanding of the underlying processes of change" was mentioned above as an objective of futures studies. This objective seems to have been given considerable emphasis in the Energy and Society project, to judge from the first part of its report which deals with past developments and in which the role of institutional conditions in the introduction of new technology is examined. The second half of the report contains a scenario of type (e), based on ideas generated by the analysis of conditions for the introduction of new technology, and representing an example of a possible chain of events which could lead to a future energy system based predominantly on nuclear power. But this scenario constitutes only a small part of the future-oriented element in the study, the main part of which can be regarded as an outline of two "boundary" developments, that is to say of type (c). As mentioned in Section 4.1, two different energy futures are presented, a solar Sweden and a nuclear Sweden. In the first case the Swedish energy system in 2015 is based entirely on solar energy, either directly or indirectly (wind power, hydroelectric power and biomass are included). In the second case Swedish energy is based entirely on nuclear reactors. These two energy developments are treated as two extreme but possible futures. Because of uncertainties and lack of knowledge we should not commit ourselves now to the choice of energy system; instead we should take active measures to keep our options open and retain our freedom of action.

The Resources and Raw Materials Study

The Resources and Raw Materials Study involved essentially a critical scrutiny of all kinds of "myths" about the future. Its task, to provide insights into the role of flexibility in the system is closely related to futures studies of type (c) in the list above (boundary developments). However, in order to obtain these insights the study group started using reference cases (type d) and critically discussed a broad range of forecasts made by others (type a). To acquire a better understanding of the system, especially in the longer term, possible alternative causal relationships were presented (e), but there was no presentation of any desirable development as such. Implicitly, however, a normative element was included in the form of a wide-ranging discussion of societal goals and of their relationships and problems, especially in connection with implementation. This also constituted a form of critical analysis of various developments suggested as desirable by other authors (type b).

The Transportation Study

As discussed in Section 4.3, the approach adopted in the Transportation Study contained normative elements, but the study itself did not include the designing of desirable futures either for Sweden or for the transport sector (or for any parts of it). Goals affected either directly or indirectly by the functioning of the transportation system in Sweden were deduced from analyses of decisions taken in the Riksdag and from government guidelines for committee work etc. The goals were applied to the identification of problems in the present-day Swedish transport system. Forecasts of a kind were made for the environment of the transport system, that is to say such aspects of society as are important to the development of transport (developments type a). In some cases alternative future developments were drawn up (developments type c). The Transportation Study also provides outlines of technological developments that might be important to the transport system, making what could be called an explorative technological forecast. It is not a normative forecast, since it does not say what technologies should, given certain goals, be developed and introduced, but it is certainly a forecast influenced by the identified goals and concerned to see how the need to limit pollution, for example, may affect the development and introduction of new technology.

In general, the Transportation Study can be characterized as an analysis of the possibilities and problems connected with a development of the Swedish transport system in a desirable direction.

## Concluding remarks

None of the three studies examined here explicitly includes predictions. However, they implicitly assume stability or continuity in many phenomena, e.g. the earth's crust, the climate, the future existence of Sweden as a nation, and the main structure of society and of international relationships. But the studies were focussed mainly on areas in which it was hoped to be able to clarify the range of probable developments, or problems and possibly desirable developments, or causal relationships.

## 5.2 FUTURES STUDIES AND DECISION-MAKING

We have already mentioned (in Chapter 2) that futures studies are intended to aid decision-making by providing information relevant in a long-range perspective, but we have also pointed out that the proximity to the application of the results may vary considerably. The results of futures studies activities will - so it is assumed - be utilized in the process of planning and policy-making, but the role of analysis in policy-making processes is always an uncertain one and in studies of complex social systems in a long-range perspective this uncertainty is even more pronounced (cf. Baehr & Wittrock, 1981). The complexity of the planning and policy-making processes makes it difficult to examine the relationship between futures studies and decision-making and the possible implications of this relationship regarding, for instance:

- criteria of adequacy and quality
- organizational aspects of futures studies, the choice of approach, and the presentation and utilization of results.

However, as regards criteria, it seems reasonable to expect that the criteria relevant to futures studies would be the same as, or similar to, those often discussed in connection with other research or professional activities which aim at providing information as a basis for public decisions (cf. Lindblom & Cohen, 1979; Rule, 1978; Whiston, 1979). In such fields as operations research, applied systems analysis, policy analysis etc, criteria of adequacy and quality have often been discussed in terms of "fallacies" (Koopmans, 1956; Hitch, 1956), "pitfalls" (Quade, 1975), "uses and abuses" (Brewer, 1975; Majone & Quade, 1980) etc. The following can illustrate the sort of criteria often cited in the literature (the related pitfalls are all quoted from Quade (1975).

- The study should address the problem relevant to the decision-makers. This may seem obvious, but:

  "A major pitfall is the failure to allocate the total time intelligently so that a sufficient share of it will be spent in deciding what the problem really is." (Quade, 1975, p 306)

  "A great pitfall of quantitative analysis and modeling is to quantify and model what we can, not what is relevant, neglecting the difficult - like Kaplan's drunk looking for his key under the street light even though he had dropped it in the dark around the corner." (Quade, 1975, p 308)

- The results should be presented at an appropriate time to match the decision-making process. This requirement is obvious in the case of any analysis geared directly to a set of decision alternatives. Those who regard futures studies mainly as forecasting may find this irrelevant. But if the role of the studies is to improve our understanding of and consequently our ability to assess, the long-term consequences of to-day's decisions, the requirement regarding timing can be seen to be highly relevant to futures studies too. In our three case studies the time element was very important, as will be seen from our later discussion of the utilization of these studies.

- The decision-making framework should be taken into account.

  "Another pitfall associated with implementation is the "myth" that there is a unique decision-maker. Analyses are ordinarily designed and carried out, although perhaps not always deliberately, as if they were to assist a solitary decision-maker who had full authority over acceptance and implementation." (Quade, 1975, p 314)

- Bias should, if possible, be avoided. This is recognized as a difficult problem:

  "The pitfall we are concerned with is honest but unconscious bias - one form is to allow oneself to be blinded by a preconceived idea as to how the analysis should come out; another is to be so caught up in an organization that one

> follows its cherished beliefs - the party line - with a mind closed to alternative approaches." (Quade, 1975, p 304)

Dror has suggested "preferization" as a primary criterion of acceptability in policy analysis (Dror, 1975):

> "Social science research is judged by accepted validity and reliability criteria. Policy studies which are oriented toward reality-improvement usually deal with subject matters for which the inherent characteristics, available research methods, time pressures, and data inaccessability do not permit findings that meet the accepted criteria of "normal" social science research. At the same time the serious implications of policy studies require some explicit and responsible acceptance criteria, inter alia, as a safeguard against essentially political and personal recommendations being presented as "science-based". - The appropriate criterion for policy studies is, in my opinion, "preferization". If the study can improve policy-results (in terms of explicated values and goals) so they will be "preferable" to those otherwise achieved, then the "preferization" test is met and the policy study should be judged acceptable".

However, futures studies are usually less closely tied to decision-making than the kind of policy studies discussed by Dror. The term futures studies is not usually applied to studies directly comparing and evaluating well-defined decision alternatives. In the field of futures studies the criteria of "preferization" can therefore hardly be made operational. Instead, the route chosen by Majone (1980) in his discussion of criteria of quality and adequacy in systems analysis seems more applicable to the field of futures studies. Majone argues that, like scientific research, applied systems analysis is essentially a craft activity, and he considers that skill in choosing and using various tools, for instance, is a major "craft" aspect in knowledge-production processes. His conclusion is that relevant criteria must deal with the process rather than the results.

To discuss in general the criteria of adequacy or quality in such a diverse field as futures studies seems hardly practical; different criteria may be relevant to different types of futures studies. It seems reasonable, for instance, that the demands on a one-month study by a consultant for a private company should differ from those attaching to a major project concerned with important societal problems and financed by a government organization or research fund. In our present discussion of futures studies in this chapter we refer mainly to the latter type of undertaking, since the three case studies described in Chapter 4 all belong more or less to that category. We could also say that our focus tallies by and large with the second of Amara's three categories (possible, probable, and preferable futures) in his typology or "summary map of the futures field" as outlined below (Amara, 1980, p 12).

SUMMARY MAP OF THE FUTURES FIELD

| | Possible | Probable | Preferable |
|---|---|---|---|
| Objectives | Open up | Analyze | Educate preferences |
| | Alert | Evaluate | Win support |
| | Stimulate | Systematize | Guide choice |
| Roles | Image-driven | Analytically-driven | Value-driven |
| Tools | Perceptual | Structural | Participative |
| Practitioners | Visionaries | Analysts | Charismatic leaders |
| | Geniuses | Methodologists | Social Reformers |
| | Writers | Futures researchers | Writers |
| Organizations | None or one-person dominated | Think tank | Advocacy group |

Amara points out that the characteristics have been (over-)simplified to create a framework and that the divisions should not be perceived as rigid and clear-cut, as in fact they are fuzzy and indistinct. In his comment on the middle category he says:

> "Futurists dealing with the probable almost always need the resources that only a team can bring to bear. In part, this is because the substantive areas span such a wide range and in part because the principal tools intrinsically rely on small group processes".
>
> "It is inevitable that the analysis of the probable has built into it some dimension of the preferable. The two components are often difficult to separate. In fact, one of the most difficult issues facing our high-technology society today is to find more effective ways of making the expertise of analysts (or technologists) available to society with as little hidden preference bias as possible".

We shall return to the problems of criteria in the following sections of this chapter in which we discuss methods and the role of values in futures studies. But first we will examine the way in which the intended relationship with decision-making influenced our three case studies, and in doing so we will focus on:

- organizational setting vis-à-vis intended users
- the presentation and utilization of result.

## Organizational setting

An organizational feature common to the three projects was the appointment of special reference groups. These reference groups can be seen as an expression of the idea that the results of the studies were intended to be useful to decision-making. However, neither the intended nor the actual roles of the reference groups vis-à-vis the project groups or vis-à-vis decision-making can be described in terms of any simple decision-making model. The choice of reference group members seems to have been influenced both by a wish to include decision-makers (or representatives of decision-makers) in the particular field as well as knowledgeable people who could assist the project groups with valuable information, views and contacts.

The organization of the Transportation Study seems to indicate a closer and clearer relationship to decision-making than the other two studies reveal. The aim of the study was to supply a background for decisions about research priorities in a long-term perspective. Since three members of the reference group were also members of the Transport Research Delegation, decision-makers seem to have been clearly represented in the reference group. In reality, however, the relationship to decision-making was rather complex. The study dealt primarily with the Swedish transportation system and with its possible future developments and problems. As was mentioned in Chapter 4, the reference group emphasized at an early stage of the project its wish that the study should be read widely, not only by transport researchers but also by planners and people taking part in the transport policy debate. This ambition influenced the orientation and presentation of the study, but the reference group did not regard it as its task to steer the study in any more direct way. Nevertheless, it acted as an important link between the project and the relevant decision-makers as the latter became familiar with the study already before final publication; this link-up seems to have been of decisive importance to the impact that the study had among other things on the new 3-year research plan of the Transport Research Delegation (which was issued before final publication of the Transportation Study).

Both the Energy and Society Study and the Resources and Raw Materials Study were attached to the Secretariat for Futures Studies, and their reference groups included people from the various ministries concerned. In spite of these similarities the projects developed rather differently when it came to their relationships with their reference groups and the balance between their two roles of serving the "centre" (to-day's policy-makers) and the "periphery" (the public debate). Two factors in particular seem to have been of decisive importance here, namely the greater political controversy attaching to the energy issues and the more provocative mode of analysis chosen by the energy project group.

Before the Energy and Society Study was launched as a project of the Secretariat for Futures Studies, it had already been started in an _ad hoc_ government committee on energy policy (_Energipolitiska delegationen_). But the political situation made it difficult to carry out this kind of study within this committee framework. The transfer of the study and its personnel from the committee to the more open-ended futures-study setting can be seen as a way of solving an awkward political question. In the earlier setting the energy group was directly involved with policy formulation for the central government. In its new setting it acquired an additional role, which became the dominant one, namely to serve the public debate. The group chose to emphasize this new role, a sign of which was its few and irregular meetings with the interministerial reference group.

The Resources and Raw Materials Study was launched directly as a project of the Secretariat of Futures Studies. It thus had no history of political controversy involving different ideas about role or focus. In this case the project group chose to remain close to the information network provided by the inter-ministerial reference group headed by the Under Secretary for Commerce.

In the case of both these projects it was clearly stated from the outset that the role of the reference group was purely advisory, that is to say it had no governmental control function. From the point of view of the project groups, this meant that certain types of influence were accepted, while others caused the groups to stress their independence. The Resources and Raw Materials group felt it was important to be independent when it came to sectoral interests and reacted against suggestions tending to limit the time perspective of the study. The energy group was obviously operating in a much stronger field of political forces, and it is not surprising that it also encountered the fiercest attack on its independence. There was one incident, which became known to the general public, when a member of the reference group attached to the Ministry of Finance tried to stop the publication of a background study (the so-called "Goodies for the Centre" debate).

To sum up: in all our three case studies, reference groups were established and intended to function as links with the decision-making processes. The actual impact on decision-making as a result of this link was probably stronger in the case of the Transportation Study than of the other two, in which the task of stimulating the public debate was given more emphasis.

## Presentation and utilization of results

At the most immediate - some would say superficial - level, the utilization of futures studies touches upon the question of the style and presentation of the results. Conceptions of style form part of the paradigmatic views of scientists and bureaucrats. But the style of an author is also influenced by his views about what audience he intends to reach.

On a more general level, three major classes of problems can arise in connection with the presentation and utilization of futures studies, namely problems arising from:

- the complexity of the processes of policy-making and planning
- the multiplicity of ways of utilizating the studies

- differences in the conception of style and the choice of quality criteria between the academic world and the bureaucracy and between different academic disciplines.

Policy-making has sometimes been seen as a process in which societal problems are identified and possible solutions to such problems are reviewed, selected and implemented. Futures studies can perhaps indicate possible developments in the environment of a given system, or in the system itself, thereby making it easier to identify potential problems. They may also contribute more directly to problem-solving by pointing out alternative courses of action open to a decision-maker. However, it is not usually possible for futures studies of large-scale social systems to focus on a single clearly identifiable policy-maker or actor. Instead the information generated will tend to be relevant to a whole network of more or less intertwined groups of policy-makers. Furthermore, these various policy-makers will often uphold different and possibly conflicting sets of goals and objectives. Hence, futures studies cannot simply serve as a basis for a well-ordered sequence of rational choices, whereby a decision-maker adapts his 'planning-object' to its 'environment' so as to satisfy a given set of goals and values. The Swedish studies presented i Chapter 4 spotlight several of the difficulties involved in linking futures studies to policy-making.

However, the complexity of the processes of planning and policy-making could be regarded in different lights in connection with the utilization of the results of futures studies. The very complexity may appear as an insurmountable barrier to the direct application of the results to policy-making, and as a contradiction of the idea of the "engineering model" according to which knowledge utilization means the direct application of specific data and findings to policy-making (Premfors, 1979, p 285). Instead the generally clarifying character of the results and their impact on problem conceptions, especially in a long-range perspective, might be emphasized. In other words instead of an "engineering model", we could speak of a kind of "enlightenment model" of utilization. But complexity could also be regarded as a challenge to the "engineering model" view, since complexity demands that special attention be devoted to the problems of linkages between futurists and policy-makers, perhaps by creating a network of reference groups composed of planners, futurists and policy-makers. In this last perspective a very general "enlightenment model" might seem lacking in stringency, ducking the issue of utilization rather than attacking it head on. From the point of view of this report these divergent opinions are important, because they carry different methodological implications.

The variety of users and of ways of using the results from a futures study was recognized by the Secretariat for Futures Studies from the outset, and was explicitly dealt with in their publication policy. Many background studies were published, and efforts were made to reach a wide readership. Both the Energy and the Resources and Raw Materials group saw it as an important function to supply the public debate with background material. At that time, however, resource issues aroused much less political heat than energy problems, which meant that the receptive mood of the readership was probably rather different. The two groups also differed in their model of analysis and presentation. The energy group emphasized the problem of choice between two extreme future options, a solar and a nuclear national energy system, and focussed on the importance of keeping the solar option open. The Resources and Raw Materials group concentrated on a critical analysis of various popular beliefs.

In the case of the Transportation Study, some version of an "enlightenment model" was probably in the minds of the reference group when the question of publication was discussed at the first meeting. The reference group was quite definite, somewhat to the surprise of the project group, that it wanted the study published as an easily readable book without appendices or special background studies. However, the project group accepted this view and organized the study and its presentation accordingly.

All three studies came up in some way against problems or criticism that arose because of various expectations which were never explicitly spelled out.

- In the energy case, certain groups were expecting the study to follow an "explorative" technological tradition of futures studies (especially as developed in the United States during the 1960s). When it turned out that the study was emphasizing a particular role of futures studies, namely to stimulate debate and, in doing so, asking a lot of rather provocative questions, some segments of the public were delighted while others were outraged. The fact that energy policy in general and nuclear power policy in particular turned into a major cause of dissent on the political front just when the group was engaged on its work, contributed to the prominence achieved by its reports and the turbulence surrounding their publication, which was on a scale seldom if ever seen before in connection with Swedish futures studies or policy-oriented research projects.
- The Resources and Raw Materials project had aroused expectations in that part of the public which wanted a programme for "the alternative movement" (including the conservationists, environmentalists etc). When the result proved to be a wide-ranging presentation of internationally available material, including a critical examination of standard clichés, at least some segments of the public felt that the product was not the one they had been expecting.

- At a very late stage in the Transportation Study (when the manuscript was ready for publication) it turned out that the administrators of the Transport Research Delegation had been assuming that the study would be published not as a book but as a report in a series they had recently started, addressed to researchers and planners in the field of transportation. An argument developed between the administration on the one hand and certain members of the reference group who were also members of the Transport Research Delegation on the other. Their conflicting views probably sprang from different ideas about the potential readership, although all they actually discussed was whether the study should be published as a book or a report.

But how were the results actually used? It is naturally hard to answer this question, or even to trace possible types of impacts. What criteria should be used? Actual influence on policy? Number of buyers or readers of the studies? In the case of the studies emanating from the Secretariat of Futures Studies, it has been estimated that about a quarter of the members of the Riksdag asked for copies of one or more of them. The Energy and Society Study sold most copies. Both this study and the Resources and Raw Materials Study were translated into English and have been reviewed internationally (see e.g. Dubos, 1981, where the Resources and Raw Material Study is compared with the Global 2000 Report to the President).

The actual impact of the studies on policy-making in Sweden was probably fairly indirect. We have not made any special attempt to trace the way they were used and can thus only guess at the type of influence they may have had. The final phase of the Energy and Society Study coincided with the work of the government Energy Commission, on which some members of the futures study group served in an expert capacity. Some of the findings and concepts developed by the futures study group were taken over by the Energy Commission, but these left no decisive mark on the major recommendations. Furthermore, the commission's voluminous report did not lead, as was intended, to a new energy-policy decision in the Riksdag in the spring of 1979, since the Harrisburg accident had occurred in the meantime and the decision to hold a referendum on nuclear power had been taken. However, the Energy and Society Study played a part in the public debate before the referendum on nuclear power in March 1980, and it probably influenced people in favour of the most anti-nuclear power alternative. But another type of possible influence should also be mentioned. The Energy Study had stressed the important role of institutional conditions in developing the energy system. Less than a year after the publication of the study a new organizational body was created - the Committee for the Development of Energy Sources other than Oil (Oljeersättningsdelegationen).

As there have been no major acute policy issues geared mainly to resource problems during the last few years, the impact of the Resources and Raw Materials Study is more difficult to trace. But the fairly wide coverage of the study makes it likely that it has exerted some influence in various indirect ways. What impact it may have on the Royal Commission on Resources and the Environment which was launched directly after the completion of the Resources and Raw Materials study (and perhaps to some degree in opposition to its outlook) remains to be seen. At least it is probably safe to say that its comprehensive supply of background material has served as a launching platform for the Commission's work.

The Transportation Study was a smaller project than the other and probably reached a much more restricted readership than either of the other two. Its results may have exerted an influence in several quarters. The 3-year transport research plan has already been mentioned. A new transport policy was adopted by the Government about a year after publication of the study, but it is difficult to know whether or not it was actually influenced by the report. Copies of one chapter of the report were distributed to regional planners by a member of the Road Administation Agency. Again, what influence this can have had is difficult to judge.

Our examples of the possible utilization of futures studies in planning and decision-making processes are, admittedly, only indicative. Nevertheless they seem to support the idea that futures studies can influence decision-making, although the relationship between the two is very complex.

## 5.3 PROBLEM FORMULATION, ACCESS TO KNOWLEDGE, AND SYNTHESIS

The impulse behind a futures study project is usually the conception of some kind of problem in the chosen field of inquiry, a problem which is probably still only vaguely defined at the time when the project group is formed. Sometimes the project group is given considerable freedom to seek out and formulate the problem or problems, as was the case in the three studies reported above. Needless to say the result of a futures study depends very much on the way the problem has been conceived. In this section we will discuss the problems involved in bringing knowledge from different fields to bear on a futures problem. More specifically, our thesis is that access to knowledge from different fields, and the way the knowledge is integrated, are crucial to results and that shortcomings can lead to risks of obtaining results that are biased, irrelevant, or misleading.

### Two models for utilizing expertise

Complex futures problems usually call for the consideration of a number of different aspects, for instance technological, economic, social, and political. This may be a reason for organizing interdisciplinary teams and trying to include people who command some knowledge in the field to be studied as well as having broad academic backgrounds. Naturally there are many different approaches to the problem of acquiring and synthesizing knowledge of various kinds. We will distinguish here between two extreme models which we can call the "methodology expert" model and the "integrating team" approach.

According to the methodology expert model the study is carried out by a methods or systems expert or by a small team mainly representing this kind of expertise, and the methodology employed defines the structure for the acquisition of expert knowledge from outside. Typical examples of this case are when the Delphi method or some mathematical model provides the basic framework for the study. When the Delphi method is employed, expert knowledge is distilled from the Delphi panel; when a mathematical model is employed, input data is often obtained from various experts outside the team.

The advantage of the methodology expert model is that it automatically provides a mechanism for synthesizing knowledge from different sources. But the choice of method is not without problems, and there is a risk that the chosen structure may lead to a one-side approach. It may also lead to an insensitive or superficial interpretation of information obtained from external sources. In Delphi studies the formulation of the questions as well as the choice of experts for the Delphi panel can lead to biased results. In the case of mathematical models one could distinguish between two types of problem: problems connected with input data and problems connected with the structure of the model. The structure is often of decisive importance to results. If aspects which are inadequately represented in the model are not allowed for in evaluating the model results, the conclusions drawn can be of dubious validity. And the way the input data is used in the model may not be compatible with the way it has been determined by some of the experts, unless the experts are all thoroughly familiar with the structure of the model. This can be characterized as a problem of inadequate synthesis. As we mentioned in the case of world models (in Section 3.4), model structures are often strongly influenced by perspectives and paradigms from different disciplins which can lead to a "paradigmatic" bias. The problem of the paradigmatic bias is that, more or less by definition, it is hard to detect. Thus the economist will tend to view all problems as purely economic problems and use economic or econometric models. The technologist will automatically tend to see the problems as concerned basically with physical resources, and so on.

According to the second model, involving the use of an integrating team, the idea is to try to include in the team highly developed skills from all fields substantively important to the subject of the study. Methodological expertise can, of course, also be included in the team and arrangements made for drawing upon outside support, for instance by commissioning special background studies. In this case, however, outside support can be given a more specific focus, since each expert in the team can be responsible for his own particular field.

This second model may seem the most appropriate one if the avoidance of paradigmatic bias and the integrating of different perspectives are desired. However, this approach too has obvious disadvantages. It requires substantical resources. The choice of team members can lead to problems of balance. Some project organizers prefer to avoid having very specialized experts on the team, as too great a degree of expert knowledge often means less interest in integrating knowledge from other fields. Moreover, specific aspects of certain perspectives might come to dominate. But even if the team members cover quite a broad spectrum of interests, there is still the problem of choosing a "centre of gravity" that reflects the problem conception of the client who has commissioned the study. If the task is to provide an overview of the future of Swedish industry, for instance, it is not enough to know whether the expert is basically a technologist or an economist; it is also important to know whether his main interest is in macro or microeconomics. Generally speaking, it is usually considered important to strike a satisfactory balance between the different skills represented in the team, but the business of actually integrating different perspectives remains problematic.

Two of our three case studies, the Energy and Society and the Resources and Raw Materials studies, come close to the second model, i.e. to the integrating team approach. The Transportation Study, on the other hand, is closer to the methodology expert model, but with certain attributes of the other as well.

In the Transportation Study the group members were specialists in long-range planning methodology and systems analysis, but they also had a general background in various areas relevant to transportation policy and research, namely transport economics, statistics and certain technical areas. Two important areas of knowledge were not represented in the group, namely economic geography and tranportation planning. Without a reference group this imbalance would probably have led to an overemphasis on economic aspects. However, two professors of economic geography were members of the reference group and one of them had carried out a background study for the project already before the team was formed. Therefore the imbalance in the expertise included in the group was not reflected - at least not obviously - in the results; but it did mean that the team members had to spend a lot of time on survey work which could probably have been done more easily by others. The methodological orientation also had its advantages: the small size of the team and the similarity of its members' backgrounds made for better cooperation; it was also possible from an early stage to concentrate on the development of the structure of the study, which was of decisive importance to the integration of different perspectives.

In the Resources and Raw Materials Study efforts were made to include experts from a variety of fields in the team. There was an emphasis on people with a background in economics and the social sciences in general (4 members including the project leader) as opposed to the natural sciences (1 from physics specialized in technology assessment). An ecologist was also recruited but, tragically, he died shortly after joining the project.

The group members had broad academic backgrounds as well as professional experience in relevant fields. There was also an interest in the problems of the developing countries, which accorded well with the guidelines for the project with their emphasis on this aspect. (The interest in developing countries stemmed more from a feeling for "social justice" rather than an interest in international trade, which could of course also have been relevant).

In the Energy and Society Study the three members of the core group had backgrounds in engineering and systems analysis or physics. Two of them had experiences of administrative rationalization and systems analysis from the National Audit Bureau and the National Defence Research Institute respectively. Later the group was enlarged so as to include two economists as full-time members.

## Problems of synthesis

In the methodological expert model the problem of synthesizing knowledge from different fields may not directly surface in the course of the study, since it is usually inherent in the structure chosen at an early stage. But when the integrating team model is used, the problem of integrating knowledge and perspectives from different fields represents one of the main methodological issues that the team has to tackle. Problems of synthesis arise at various levels. At a lower level there is a need for surveys and background studies in different fields. At a higher level there is the problem of dealing with developments in the future in a cohesive way. Syntheses become difficult, particularly at this level, because of the different approaches and perspectives of the people involved. Such differences can stem from academic background or they may be related to professional experience, ideology, personality type, etc.

have been developed according to which several groups - each internally homogeneous - work on the same problem (Mitroff et al, 1977).

No discussion of the desirability of attaining a balanced view would be complete without some comment on the problems that can arise if such an ideal is pushed to the extreme. It is reasonable to suppose that the inclusion of more than one viewpoint would help to give the reader greater insight into a complex problem area. However, some perspectives may be more relevant than others to certain problems. And obviously there is no such thing as the balanced, definitive and overall perspective. In regard to certain of the purposes of futures studies, such an aspiration might even be counterproductive.

In the Resources and Raw Materials Study the writing of the final report involved a high level of integration. The report was produced jointly by the five members of the core group in a process whereby each chapter was rewritten at least once by some other member. Nevertheless the study seems to reflect the "centre of gravity" of the team members' backgrounds, in the sense that there is more emphasis on economic viewpoints and less on ecological or more natural-science-orientated approaches.

As we have seen, two economists joined the original energy group at a later stage. This led to an intensive exchange of ideas about the focus of the study and the nature of "the energy problem". But it also triggered off considerable disagreement within the group; eventually, the two economists left the group and did not take part in writing the final report, "Solar versus Nuclear - Choosing Energy Futures". In this report there is an emphasis on technical and organizational aspects of the energy problem. It includes economic considerations, but no in-depth economic analysis. An important feature of the energy project was that outside experts made major contributions by producing background papers on a series of social, political and technical

The outlook attaching to the different academic disciplines makes itself felt in a variety of ways. It may be a question of emphasis, perhaps with regard to the level of aggregation or the time perspective. A sociologist and an economist will look at people and decision-making from different viewpoints. An ecologist concerned about the flow of nutrients and the need to sustain biological systems will choose other major explanatory variables than those which an economist would choose, with his preference for analysis in terms of market forces. And certainly a technologist and a biologist will differ in their views of the role of the organic as opposed to the inorganic part of the world. Nor would they be likely, in general, to have the same opinion on various other questions such as the ability of new technology to solve tomorrow's problems.

Some disciplines are more theory-orientated than others; and in many cases there is tension within the discipline between comparatively theoretical and comparatively empirical lines. Attitudes regarding the importance of quantification can also vary. Some people prefer conceptual models covering broader areas and relationships, while others want to limit a study to what can be quantified. There may also be fundamentally different assumptions regarding causal relationships, as for instance between more or less deterministic modes of interpretation of social phenomena.

It could be argued that in order to obtain a balanced synthesis, attempts should be made to include in the team representatives of a multitude of different backgrounds, viewpoints and attitudes in a variety of dimensions. On the other hand the larger and more heterogeneous the team, the more difficult becomes any real integration of different perspectives (cf. Rossini & Porter, 1979). The team may break down altogether, or the study might lead to a number of different publications or sub-results without any integrating framework. In fact, group work methodologies issues.

To succeed in fully integrating a number of essentially different perspectives in a complex futures study is perhaps hardly feasible. Some kind of paradigmatic bias is almost certainly unavoidable. The important point is rather to convey to the reader an understanding of what the study covers, so as to avoid misinterpretation of the results.

Thus, the outcome of a study often reflects in many ways the procedure by which it was set up and the various professional backgrounds and academic disciplines to which the members of the project group belong. Bias stemming from professional background may be at least as important a problem as the kind of bias which value preferences of a political or any other kind may engender (cf. 5.5).

## 5.4 SCENARIOS AND MODELS - SOME CHOICES AND CRITERIA

In this section we shall discuss some issues raised in the methodological debate in connection with scenarios and models. But first some comments on tools in general.

Criticism of the tools of the futurist is not uncommon. Sometimes the criticism concerns a specific technique, for instance Delphi (cf. Section 3.1) or mathematical modelling (Clark & Cole, 1975), and sometimes a set of tools and techniques (Hoos, 1978, pp. 53-66). But even the whole field of study has been called in question in light of an examination of some of its tools (Asplund, 1979).

It cannot be denied that the tools have major weaknesses, that they have sometimes been misused or misunderstood, and that inaccurate claims have been made regarding results. However, we would argue on various grounds that an examination of specific uses and interpretations of some tools is insufficient for an assessment of the field of futures studies as a whole.

First, there are many variants of the different tools, and many ways in which they can be adapted to the problems at hand. This point has been discussed with respect to Delphi in Linstone (1978). The way in which the tools are used is also pertinent. The role of craftsmanship in the use of scientific tools has been examined, for example, by Ravetz (1971) who emphasizes the importance of craftsmanship for the validity of research results. And craftsmanship of this kind is naturally even more important in a young field such as futures studies.

Secondly, an assessment of the field of futures studies cannot be based solely on separate examinations of the potential of each individual tool since different tools are sometimes used in combination and this can serve to reduce the effects of the weaknesses of any one of them. In the three case studies presented in Chapter 4, a combination of various tools and approaches was in fact used, including various research methods - statistical analysis to mention but one - long since well established in other fields. Indeed it seems not uncommon to find quite traditional research methods being used exclusively or almost exclusively in futures studies, for instance in examining the dynamics of change in a social system. This naturally only holds if we regard the field of futures studies as determined by its focus on a futures problem rather than by its choice of method.

We shall limit our discussion here to scenarios and mathematical models, partly because these are both among the methods most frequently used in futures studies, and partly because they were both employed in our case studies.

Issues that have been raised in the methodological debate include the relative advantages of models and scenarios. For instance, it has been asserted that:

i) scenarios are superior to models (Julien et al, 1975, pp.51-52)

ii) mathematical modelling is superior to other approaches (Meadows et al., 1972)

iii) it is important to develop criteria of adequacy for the development and utilization of scenarios (cf. e.g. Julien et al., 1975).

As to the third point, we concur; we shall comment on it in some detail below. First, however, we will examine the first two assertions, which we regard as being based on misconceptions and in need of some modification; we go so far as to designate them "methodological misquestions".

Scenarios superior to models?

As we have seen (Section 3.3) the term scenario as used in the futures field has a number of different connotations. Thus if we want to compare "the scenario method" with other methods, the implicit methodological content has to be specified. In Chapter 3 we identified a number of methodological ideas, which may be in people's minds when the scenario method is being discussed. Examples are:

- We cannot predict the future, but by designing a number of plausible and consistent descriptions of hypothetical future developments or situations (scenarios) we can delimit the uncertainty space which we want to consider in studying the problem at hand.

- If we want to anticipate problems and improve our future situation we should not simply assume that present trends will continue; instead we should design desirable futures (scenarios) and identify branching-points and mechanisms of change, so as to find actions which can affect developments in a desirable direction.

- It may be useful to try to synthesize fragmented, dispersed and sometimes vague knowledge into a holistic and consistent picture (scenario) of a future development or situation.

Each of the ideas mentioned above applies to one or more of our three case studies. In all of them "the scenario method" was used, although in the case of the Resources and Raw Materials Study the scenarios developed were not used directly in the final report, and in the Energy and Society Study the group itself did not use the actual term scenario. However, comparing the three studies, we find that in so far as scenarios occurred they were very different both in form and in the role they played.

As to the form of a scenario, it may perhaps be stated that it should be a verbal description of a future course of events, for instance with the emphasis on causal relationships. The role of a scenario may be to indicate a probable development, or a desirable development, or perhaps a "surprise-free" development (as a reference projection). And the role of a set of scenarios, describing possible but extreme developments, may be to delimit an uncertainty space. One difficulty is that the use of "the scenario method" sometimes but not always means that the scenarios have to meet specific requirements as to both form and role. However, if role requirements only are considered important, then there is a certain overlap in meaning between "the scenario method" and other methods as futures with specific roles, e.g. futures representing probable or desireable developments, can also be constructed with the help of Delphi, forecasting techniques, mathematical modelling, etc. Naturally this complicates comparisons of "the scenario method" with other methods used in futures studies.

Let us examine how this comparison has been made in the study "La Mēthode des Scēnarios" (Julien et al., 1975) where the authors argue that the scenario method is superior to "modelling" (pp. 51-52). Modelling here refers to the development and use of formal, quantitative models. A distinction is made between two types of scenarios with different roles; one is to describe a desirable development ("scēnario d'anticipation"), the other to describe a probable development (scēnario exploratoire). A comparison is then made with normative models (modèles normatifs ou modèles de design) and with forecasting models (modelès predictifs) respectively.

Models are heavily criticized for a number of reasons including inadequate social theories, the risk of oversimplification, difficulties in evaluation, etc. And the authors emphasize that the less the model is based on a valid theory, the more it becomes a source of errors. Thus "modelling becomes dangerous and must be abandoned".

We can agree with this criticism for cases in which models are used for the long-term forecasting of social systems. However, when theories are lacking it is not necessarily easier to use scenarios than to use models for forecasting. And for cases in which we want to outline a desirable development for a complex system, the main problem is not necessarily the inadequacy of existing theories. So the arguments[1)] for the superiority of the scenario method to modelling for the design of desirable developments, seem insufficient. Actually the negative attitude towards the use of models seems to be due to the fact that the authors only consider two possible purposes for using models in futures studies: to produce forecasts or to outline desirable developments. We mentioned in Chapter 3 (p. 42) that mathematical models in futures studies are often used as "what-if" models or learning models. They may, for instance, be used to produce estimates which are intended only to be of the right order of magnitude, or to illustrate the difference resulting from a change in a particular parameter. They can thus be used to supply insight into the problem studied. In fact, this was the kind of way in which models were used in the Resources and Raw Materials Study and the Transportation Study.

In the Resources and Raw Materials Study it was decided at a fairly late stage to use a model, or rather a system of models, in order to examine the future need for various raw materials

---

1) We refer here to the following arguments: "Il n'existe pas encore, pour les systèmes complexes, de modèles normatifs véritables; ce type de modèle pose des problèmes tels qu'il faudra de longs efforts de recherche théorique et de modélisation avant qu'ils puissent être résolus de façon concluante. En attendant, la méthode des scénario, comme nous allons tenter de le démontrer, peut être utilisée avec des résultats satisfaisants."

in Sweden. The results were not expected to be accurate in the long term, because of the limitations involved in dealing realistically with substitution and technological changes in an input-output model. The use of the model was restricted to the next 10-15 years. It was not the absolute figures resulting from the runs of the models that were regarded as important, however, but the variation between the results of the runs that were based on different assumptions about the general situation in the world ("conflict between developing and industrialized countries" etc).

In the Transportation Study the same econometric model system as in the Resources and Raw Materials Study was used. However, no new model runs were made, the results from some earlier runs being used after certain interpolations had been performed. The results were used to give a quantitative example of the general economic development in Sweden in each of the three scenarios (cf. Section 4.3). Part of the information on which the scenarios were based was thus obtained from a model. In the Resource and Raw Materials Study it was the other way around; part of the input data used in the model runs was based on scenarios. This way of using scenario-model "hybrids" is probably not unusual. For instance scenarios were also used in combination with models in the interfutures project (OECD, 1979).

Thus, the question as to whether or not scenarios are superior to models seems to be something of a "misquestion". In fact it seems to follow from our illustrations and our examination of the different connotations of the terms, that they should be regarded rather as complementary methods. More specifically, it should be noted that:

- mathematical models can be used to provide information which can then be used in the construction of scenarios (and vice versa)
- scenarios and mathematical models can be used independently in the same project
- sometimes the term scenario is used to designate a specific set of paramters or run of a computer model.

## The superiority of modelling?

Now let us turn to the second of the methodological assertions mentioned above, namely that mathematical modelling is superior to other approaches in futures studies. More specifically we refer here to "Limits to Growth," in which the authors argue that their model, which is of the systems-dynamic type, has advantages over mental models because it is a formal or mathematical model (Meadows et al., 1972, p. 22). As there is no comparison with other approaches such as Delphi or scenarios, the reader gets the impression that formal models and mental models are the two sole possibilities between which a choice has to be made.

We are not going to try to summarize here the intensive debate that followed the publication of "Limits to Growth." The model, as well as the rather farreaching conclusions drawn from it, have been extensively criticized. In-depth studies of the model were made at an early stage by the Science Policy Research Unit of Sussex University (Cole et al., 1973). A number of other global models have been developed which in varying degrees have produced different results. Interesting comparative evaluations of the various global models have also been made (Clark & Cole, 1975; Cole, 1977).

If we apply the distinctions made in Chapter 3 between different purposes for the use of mathematical models in futures studies, the "Limits to Growth" model as well as most other formal global models prove not to be forecasting models at all, they are, rather, some kind of "what-if" models. Such models can be said to produce inputs to the mental model of the user of the results. In other words some kind of mental model is ultimately used; it has not been replaced by a mathematical model. It could of course be argued that the development and use of mathematical models always means an improvement in the information available. In our view this is not necessarily so. If the assumptions on which the model is based are not sufficiently well understood, the results may be so misleading that the user would have been better off without the model.

Criteria of adequacy

In the introduction to "La Méthode des Scénarios", a Canadian research project, the importance is emphasized of developing criteria for the construction and utilization of scenarios (Julien et al., 1975). The study also makes some interesting contributions in this context itself. Apart from elaborating on such common criteria as coherence, consistency, plausibility etc, it compares and evaluates various scenarios and conceptions of the scenario method. From a comparison of three approaches (Kahn, DATAR, Ozbekhan), four aspects of general importance in scenario-writing and futures studies emerge:

- the role of values
- the view on past-present-future, e.g. the use of historical analogies
- treatment of decision-making
- causality.

However, the discussion in the Canadian report does not seem to be directly applicable to our three case studies. The structures of the scenarios in Energy and Society and the Transportation Study are quite different from those described in the Canadian project. The scenarios in the Resources and Raw Materials project are more in accordance with the criteria discussed but, as we mentioned above, these scenarios were not directly used in the final project report.

There are indications that a good deal of effort has been devoted to the development of the scenario method during the last 15 years. In any major futures studies considerable interest seems to be devoted to methodological issues and criteria of adequacy are often developed. This was the case in the Resources and Raw Materials project as well as in connection with scenario work in Swedish defence, for instance, and in the DATAR project in France. But there are still no obvious signs of the emergence of any generally accepted criteria.

One reason for the diverging views on scenarios could be related to the existence of different futures studies doctrines. Hannigan has pointed out the coexistence of two schools of thought, a more deterministic "technocratic futurology" based on an American tradition, and an explicitly normative school based on a European tradition (Hannigan, 1980, p. 321). Technocratic futurology or "technocratic historicism" is described as resting "on a set of theoretical assumptions derived from the evolutionary model of a post-industrial society as suggested by Daniel Bell and Herman Kahn" and concentrating on "specific economic and technological forecasts".

It seems obvious that different ideas about the extent to which the future is predetermined will affect scenario-writing. For instance, scenarios made for land-use development in France, with its system of centralized planning, will differ from U.S. macro-scenarios based on the assumption of non-intervention on the part of the Federal Government. One might conclude from this that scenarios, and futures studies in general, should be presented together with information about the author's opinion of our capacity to "create the future".

However, the differences in the scenarios in our three case studies do not seem to be attributable to the two doctrines mentioned above. Both Energy and Society and the Transportation Study can be said to belong to the European school of thought, and yet there are noticeable differences between their scenarios. One reason for this is that in the energy case the scenarios describe two possible but extreme developments of the planning object, while in the Transportation Study the purpose of the scenrios is to describe the development of the environment of the planning object. Thus further developments of the scenario method seem to require that a certain number of disctinctions be made, e.g. between scenarios that describe the planning object itself and other scenarios, and between scenarios concerned with planning objects which are centrally controlled and others concerned with decentralized decision-making, and so on.

The dissimilarities between the scenarios in our three case studies make it difficult to draw any general conclusions about criteria of adequacy. But we will discuss here one problem which seems relevant to all three studies, and indeed to a number of others as well. The scenario method usually involves the use of several scenarios, and our problem concerns the choice of parameter (or parameters) to be varied between them. This choice is not without its problems, and it can be of decisive importance to the study.

There are a great many kinds of possible parameters, e.g.

- economic (low or high economic growth; protectionistic or free world trade, etc)
- technological (type of energy carrier used as supply; optimistic or pessimistic assumptions about new technologies, etc)
- elements of national policy (length of working week; limitations on the use of nuclear power, etc)
- social (new life-styles, etc)

There may be grounds for varying several parameters between the scenarios. But the number of combinations soon grows with a growing number of parameters, and for practical reasons the number of scenarios - and consequently the number of parameters - must be kept down. As few as 1-2 parameters and about 3-6 scenarios seems ususal in practice (Svedin, 1979; Wilson, 1978). The choice of parameter or parameters is thus of crucial importance, particularly when only one is being used. Nevertheless this choice is rarely discussed at all; it seems to be made automatically, at least in non-interdisciplinary project groups. As we have mentioned earlier (3.4, 5.3), the academic backgrounds of project members may have a decisive effect on the way in which the structures of real world problems and developments are perceived. In scenario-writing these affiliations may easily determine the choice of, let us say, a technological, an economic, or a social parameter. The choice of parameter can result in the scenarios giving a misleading impression. If, for instance, the task of the set of scenarios is to delimit a space of probable developments, and if an important but very uncertain parameter is not varied, there is a risk of bias which could sometimes be designated as a kind of paradigmatic bias (cf. 5.3).

There are special problems connected with the choice of parameter for a set of scenarios, when these represent alternative developments of a comparatively large system, e.g. different global developments (macro-scenarios), and the function of the set of scenarios is to provide a background for studying alternative developments in a sub-system (micro-scenarios). If the macro-scenarios are to serve a useful purpose, it must be possible to trace the impact of the difference between them on the micro-level. Difficulties in establishing this kind of linkage were reported from the Resources and Raw Material Study.

The great diversity of forms and roles that scenarios can assume, and of the areas in which they can be applied, makes it unlikely that future developments of the scenario method will lead to generally accepted rules for their construction or criteria of their adequacy. Further developments may instead give us a better typology of scenarios and guidelines for avoiding or detecting bias.

As an example of the difficulty involved in developing general guidelines for scenario-writing, we can conclude by listing some "criteria of adequacy" based on an analogy with photography. The analogy was developed with scenarios in mind of the kind tried in the Resources and Raw Materials Study. The "rules" may seem self-evident, but the extent to which they can be given relevant interpretations will vary with the form and role of the scenarios, with the nature of the problem area, etc. More importantly, they do not necessarily cover all requirments (e.g. there are no guidelines concerning the choice of parameters for a set of scenarios), and a negative effect of applying a predetermined set of rules can be that attention is diverted from possibly more important requirements. Nevertheless we think the list may be fairly inclusive for certian types of scenarios and that it also throws some light on problems which are general in the field of futures studies, i.e. exist also when other tools are being used.

An example by analogy

Let us recall how an ordinary photograph is made and the obvious criteria we pose for the quality of our snap-shot.

- The picture must be appropriately edited. If you are supposed to be taking a picture of a person, you should probably concentrate on the face. In particular the photographer should not cut off his sitter's head.
- The picture should not be out of focus. The choice of framework has to be appropriate. And you have to close in on the subject with the right depth, neither too superficially nor too deeply. In scenario terms your choice of causal explanation must be on the relevant level. If you are asked for a scenario on "the development of the growing importance of regional independence",

the choice of the "existence of certain charismatic personalities" as an explanation may capture some of the flavour of the problem but probably misses the main target.

- The picture has to reflect movement correctly. If the subject moves a good deal during the "exposure time", i.e. in relation to the methods of investigation chosen, then the job is badly done. If, for example, you are supposed to be saying something about the year 1990 in a quickly expanding field like microelectronics, and you only have data points from 1950 and 1960, you will probably be in trouble. Or if you intend to outline the future of the world economy in the year 2050 and you base your forecast on an econometric model of world economies of today without provision for any change in technology, you have probably made the same type of blunder.

- The picture should be correctly exposed. If it is underexposed there will be too little light on the theme you are supposed to be illuminating. Some objects are dark in themselves, but we are not talking about that here. But there are some studies suggesting a number of interesting possibilities which have not been used or which have not been pursued as far as would have been possible. Common reasons for this kind of error are lack of time, lack of knowledge of the possibilities, or "sloppy work" in general. The overexposed picture represents an interesting fault. Normally there is not much harm in overdoing things, except that it is a waste of time. In the case of futures studies, however, there is a special type of "overexposure" that should be guarded against. It is connected with an exaggerated claim to clarity and depth in the picture presented, based on the amount of work done. A very detailed computer model need not be particularly relevant to the problem it is supposed to deal with, just because it is detailed. But such a claim is all too easily made.

A scenario might be written in such a way as to be full of details while at the same time concealing most of the problems connected with the choice of appropriate variables or causal relationships, i.e. it is overexposed.

- The picture should be on good-quality film. If the "substance" on which the picture is to be fixed is poor, the picture may also be poor. Pre-existing ideas about points of interest and relationships (e.g. an ideology or method of scientific approach) might represent such an "old film". The ideas may be outdated in relation to the problem.

- The film should not have been damaged by light or heat in the camera. If a research group is exposed to various kind of pressure (e.g. political, religious) while formulating its picture of the future, the deformation of the final picture can be regarded as the analogue of a distortion on the film when it is in the camera. A good research group takes steps not to prevent the "heat" (because that is not controllable by the group), but to see that its presence does not have too much affect on the picture formation.

## 5.5 VALUES AND THE STUDY OF THE FUTURE

The study of the future - it is often argued - can never be a dispassionate search for eternal truths. Instead it is a field fraught with intense emotions where diverse interests and values clash. It is quite commonplace nowadays to hear futures studies depicted in terms like these. Such remarks are certainly not unsubstantiated, but they are often ambiguous. Exactly what kind of values are supposed to be the crucial ones? And why should their role in futures studies be different from their role in, for instance, political science?

Sometimes the answers to such questions are thought to be so obvious as not to need spelling out: after all, since the future does not exist, we are dealing not with hard facts but with ideas, hopes and dreams. Thus futures studies differ inherently in this way from those social sciences which can produce empirical knowledge of the real world.

We have already tried to show that such a position suffers from some obvious weaknesses. Futures studies do not deal exclusively with future events. And when they do, the emphasis is seldom upon scientific prediction. So the suggestion that by reason of their very object of study futures studies are barred from an analysis of the existing world is, at best, only partially correct. Nor is it clear why knowledge should be limited to phenomena that we can experience in varying degrees, because they exist here and now. Such a narrow empiricist stance would cause problems not only for futurists, but also for historians or for any scholars engaged in the more theoretically developed branches of the natural sciences.

Furthermore, in several of the social sciences - economics is an example - scholars engage in simulation modelling, for instance, without worrying too much about whether their activities might violate some narrowly conceived empiricist position. Thus the problems that face futurists in terms of their object of inquiry are not necessarily fundamentally different from the problems with which scholars in many other disciplines have long been familiar.

In this section, however, we shall focus on the role of values in futures studies. In Sweden this role was seen as a crucial one already at the outset of the discussion about the organization of futures studies. Thus, in setting up a government commission of futures studies in 1971, Olof Palme, then Prime Minister, explicitly stated certain normative principles for official Swedish futures studies:

> "An important and necessary requirement is that we ourselves should study the future, for ourselves and for our world, and that this study should be based on democratic objectives and express demands for international solidarity."

Similar phrases echoed in the report of this commission:

> "Future-creating activity must contain the following important elements: a strengthening of the democratic process, more exchange of information between different actors, and guarantees of participation by citizens in decision-making... It follows that a better understanding of how values change and of how such changes find expression in planning decisions, assumes a position of central importance in the futures studies picture." (To Choose a Future, 1974, p. 10, 15, 26)

In a recent survey of "the futures field" Roy Amara, distinguishes between some main types of futures studies pertaining to the probable, the possible and preferable. However, he also specifies a few general criteria that he deems to be applicable to the whole futures field. And one of these criteria is value explicitness:

> "Thus, whatever the premises revealed and objectives identified, it is is important to clear the air by describing as precisely as possible values relevant to the output produced. For example, if we're making a forecast on the likelihood of economic growth, it would be imperative to know whether we are low- or high-growth advocates..." (Amara, 1980, p. 21)

Similarly, in the American National Science Foundation report on "The Study of the Future: An Agenda for Research" Willis Harman comes out strongly in favour of value explicitness:

> "Perphaps there are only two kinds of research: overtly normative and covertly normative... There needs to be an emphasis in futures research which is explicitly normative, which in the best tradition of Utopian thinking attempts to delineate an image of a workable and desirable midterm future that we can get to from here... The more difficult are times in the near-term future, the more crucial will be a vision of what is worth building and where is worth going." (Harman, 1977, p. 76 and 84)

Thus, the need for value explicitness has been a recurring theme in both Swedish and international literature on futures studies. However, the values involved may be of different types. Three broad types can be distinguished: political, professional and paradigmatic. The discussion on the role of values in futures studies has often focussed on values of a tangibly political or partisan nature. The idea of value explicitness then translates into a two-pronged demand:

- political values that, somehow, underlie or condition the way in which alternatives are outlined, forecasts and scenarios are made, shall - to quote a programmatic statement of the Swedish Secretariat for Futures Studies (Swedish Secretariat, 1975, p. 16) - "be reported openly".
- outlines of alternative development paths and futures must include a description of the way previously specified values can or cannot be satisfied in the various alternatives. Or to cite the Swedish Secretariat's programme: "how discrete

valuing dimensions can be met alternative by alternative". (Ibid.)

Sometimes, as in Roy Amara's statement quoted above, these two demands merge into one. If some politically controversial value is included in the forecast, economic growth perhaps, then it is safe to assume that the forecaster's attitude to that value might condition the forecast. Hence, it should be clearly spelt out in the study itself. In the case of the Swedish Secretariat for Futures Studies, some major value dimensions, common to all the Secretariat's projects and work, were formulated: international solidarity, justice between different groups in the population and between different generations, the view of human beings as active and creative, and, finally, political participation, involving a commitment "to broaden and deepen democracy in our society". These value dimensions were taken as a point of departure for a number of discussions within the different project groups set up by the Secretariat. In the two projects reported here, the group studying resources and raw materials spent a considerable amount of time trying to elucidate more precise meanings and implications of the values indicated, whereas the energy group devoted little or no time to such ends. However, neither of these groups, nor the other two involved in the first generation of the Secretariat's projects, came anywhere near to satisfying the kind of criterion that Roy Amara had in mind. On the contrary, the energy group made a virtue of positing freedom of action in the choice of a future energy system as a paramount objective. And the group certainly did not state explicitly whether it preferred the nuclear or the solar option or any particular combination of the two.

It is important to note, however, that ideas about political and societal values can play a more indirect but potentially more important role in futures studies than appears from the traditional demands for value explicitness. The way in which we value a particular alternative future will affect our decision to include it among the alternatives to be seriously considered. As Göran Hermerén among others has argued, we might very well include an alternative in our set, although it has a very low probability. It might nevertheless be relevant because its consequences, if it were realized, would be on a catastrophic scale. So the way in which we value an alternative is, indirectly, bound to influence the conduct of a futures study (cf. Hermerén, 1976, p. 177 f). Conversely, an outline of a possible chain of future events can function as an articulation and specification of a set of values. And if such chains are really sketched with a mind open to unexpected and possibly uncomfortable consequences, they might well serve to promote a reconsideration of some of our cherished values. If for no other reason than because some of the values might prove to be in conflict in certain situations (cf. ibid., p. 183).

Political values are also revevant to the actual process of carrying out a futures study and thus pose problems of project management. How should the various ideological predilections of the members of a futures studies group be handled? How should project groups be composed with respect to the span of opinions contained within any one project group? To put it bluntly, there seems to be no sort of consensus among the authors who have treated this issue. Some people emphasize the vital importance of making sure that "forecasting groups are balanced for the ideological differences a series of our studies have shown to be important factors: liberals and conservatives, younger and older generations factions, activists and passivists, extremists and moderates, leaders and followers, and/or tough and tenderminded" (Loye, 1980, p. 101). But others claim that project groups should not be composed of representatives of widely different political persuasions and personality types. Indeed, it is sometimes argued that the production of alternative future scenarios by groups that are internally cohesive but ideologically far removed from one another is a way of getting insight into the range of real uncertainty (cf. Cole, 1976, p. 314; Clark & Cole, 1975, p. 114).

Obviously the studies discussed in our present report can yield little information that might settle this issue. However, Sweden's political culture stresses the importance of broad-based consultation and of the participation of interested parties and groups in the policy-making processes. And the consensus on the various recommendations that have been made and implemented is in itself a cherished value in this culture. So there seems to be little reason for any extra effort on top of the normal operations of Swedish policy-making to ensure this particular type of balance in the conduct of futures studies. Neither excessive imagination nor an addiction to wild alternative solutions has ever been a common problem in Swedish policy-making.

But political and partisan values are not the only values relevant to futures studies. Before a futures study can even get off the ground, those involved must have some idea as to their own professional role, i.e. they must be committed to some form of professional values. Do futurists value the descriptive and analytical type of work that is typically performed by the university scholar in the social sciences and the humanities? Or do they see themselves as change agents, or as the advocates of unconventional ideas in a bureaucratic environment?

In any real-life situation the results of a futures study will tend to contain a subtle mixture of descriptive, prescriptive and persuasive elements. And futurists are seldom pure university scholars, or pure policy advisors and consultants or pure exponents of a particular policy. Nevertheless project groups differ in the way they emphasize these elements. Both the Energy and Society and the Resources and Raw Materials studies were basically orientated towards policy-making and the public debate on policy. In neither project did any member who identified with the image of the university scholar detached from the urgency and turbulence of the political scene play more than a marginal role. Incidentally, the reverse was true of at least one of the other two first-generation projects of the Secretariat for Futures Studies, namely the project on Sweden in World Society. This project was based in a typically academic environment at the University of Lund.

The Transportation Study was broadly speaking of the same type as the Energy and Society and the Resources and Raw Materials studies in terms of professional values. But being a smaller project, it did not initiate a series of background studies on specific themes. In the Energy and Society and the Resources and Raw Materials projects a large number of people with widely differing professional values could contribute to the general work of the projects without any need for the core groups - or the interministerial reference groups - to commit themselves to the theses or values of the individual background reports.

Finally, the outcome of a futures study will depend crucially on the paradigmatic values embraced by those making the study. This term is used here not only to denote methodological principles and criteria of quality such as precision, predictability or reproducibility. Included are also ontological and theoretical assumptions: What does the reality that we are studying consist of? Are we dealing with individual human beings who are carefully calculating the costs and benefits of various courses of action? Or do we face a social scene populated by social classes involved in a struggle with one another? Or is the focus on chains of raw materials or the elements of a fragile ecological balance?

It is all too obvious that answers to such questions might condition the whole approach of a futures study project. Furthermore, the answers are often only implicit. The very choice of project members will inevitably impose a number of limitations on the conduct of a project. Outside competence can of course be tapped, perhaps for background studies or through reference groups, but the general structuring of futures problems must be carried out by the project group itself.

Of our three studies, the Transportation project group chose an approach in which planning-theory concepts, such as the distinction between planning object and environment, were crucial. To be sure, concepts of this kind formed part only of a general framework that evolved in a process of continual revision - a framework that permitted economic and geographical aspects, for example, to be included in the picture. But the general approach was clearly different from, say, a more traditionally economic one, in which the main problems often appear to be concerned with satisfying a future demand for transport under certain economic restrictions and often with no regard for the negative side-effects that transport can involve.

In this respect the Transportation Study and Energy and Society were similar. The Energy Study did not disregard the economic feasibility of Nuclear Sweden or Solar Sweden. But production costs were not seen as the determinants of the emergence of different energy systems. Instead they were regarded as being dependent upon certain more general features of social organization which, so the group argued, could be politically controlled at least in an indirect fashion and in a long-term perspective. As it turned out, conflicts of a paradigmatic nature stimulated the process of problem formulation and also triggered off an eventual rupture in the group whereby the two economists withdrew.

The Resources and Raw Materials Study group tried consistently and strenuously to adopt three basic perspectives: an economic-political view, a technological view and an ecological view.

This pluralist approach permitted the group to encompass results from a broad range of background studies and to draw upon competence from a variety of different disciplines. But it also left some disturbing problems unresolved about the more precise relationships between the three perspectives employed. However, the experiences from both the Energy and the Resources studies indicate that the cleavages that appeared in the course of the projects tended to be related to values of a paradigmatic and professional nature rather than to values of a more immediately political character.

## REFERENCES

AMARA, R., "Probing the Future" in J. Fowles (ed) Handbook of Futures Research (Greenwood Press, Westport, Connecticut, 1978).

AMARA, R., The Futures Field (Institute for the Future, Menlo Park, California, Sept 1980).

ASPLUND, J., Teorier om framtiden (Liber, Stockholm, 1979).

AYRES, R., Uncertain Futures - Challenges for Decision-Makers (John Wiley & Sons, New York, 1979).

BAEHR, P.R. & B. WITTROCK (Eds), Policy Analysis and Policy Innovation (Sage Publications, London and Beverly Hills, 1981).

BREWER, G., An Analyst's view of the Uses and Abuses of Modelling for Decisionmaking, Rand p-5398 (Rand Corporation, Santa Monica 1975)

CLARK, J. & S. COLE, Global Simulation Models (John Wiley Interscience, London, 1975).

COLE, S. et al, Models of Doom: A Critique of "The Limits to Growth" (Universe Books, New York, 1973).

COLE, S., Global Models and the International Economic Order (Pergamon Press, Oxford, 1977).

CORNISH, E., The Study of the Future, World Future Society (Washington, D.C. 20014, U.S.A., 1977).

DROR, Y., "Some Fundamental Philosophical, Psychological and Intellectual Assumptions of Futures Studies" in C.H. Waddington (ed) The Future as an Academic Discipline (Elsevier, Amsterdam, 1975a).

DROR, Y., "Some Features of a Meta-Model for Policy Studies", *Policy Studies Journal* 3 (1975b) 3, pp 247-255.

DROR, Y., "Think Tanks: A New Invention in Government", pp. 139-152 in C.H. Weiss and A.H. Barton (eds), Making Bureaucracies Work (Sage Publications, London and Beverly Hills, 1980).

DUBOS, R., "Half-Truths About the Future", *Wall Street Journal* 1981-05-08.

HANNIGAN, J., "Fragmentation in Science: The Case of Futurology", *Sociological Review*, Vol. 28, No 2, 1980.

HARMAN, W.W., "On Normative Forecasting" pp. 76-85 in W.I. Boucher (ed) The Study of the Future: An Agenda for Research (National Science Foundation, Washington, D.C., 1977).

HERMEREN, G., "Rational Arguments - On the Role of Norms and Values in Futures Studies" in S. Schwarz (ed) Knowledge and Concepts in Futures Studies (Westview Press, Boulder, Colorado, 1976),

HITCH, C., "Professor Koopmans on fallacies: A comment", *Operations Research*, Aug 1956, 4 (4).

HOOS, I., "Methodological Shortcomings in Futures Research" in J. Fowles (ed), Handbook of Futures Research (Greenwood Press, Westport, Connecticut; London, England, 1978).

JULIEN, P-A., P. LAMONDE, D. LATOUCHE, La Méthode des Scénarios (La Documentation Française, Paris, 1975).

KOOPMANS, B., "Fallacies in Operations Research", *Operations Research*, Aug 1956, 4 (4).

LINDBOM, C.E., & COHEN, D.K., Usable Knowledge (Yale University, Paris, New Haven and London, 1979).

LINSTONE, H., "The Delphi Technique" in J. Fowles (ed), Handbook of Futures Research (Greenwood Press, Westport Connecticut; London, England, 1978).

LOYE, D., "Personality and Prediction", *Technological Forecasting and Social Change*, Vol. 16, 1980, pp. 93-104.

MARTINO. J., Technological Forecasting for Decisionmaking. (American Elsevier, New York, 1972).

McHale, J., "An Assessment of Futures Studies Worldwide", *Futures*, Volume 8, Number 2, April 1976, pp. 135-145.

MAJONE, G., The Craft of Applied Systems Analysis, WP-80-73, IIASA (A-2361 Laxenburg, Austria, 1980).

MAJONE, G., & QUADE, E., Pitfalls of Analysis, (Wiley, London, 1980).

MEADOWS, D., et al, The Limits to Growth (Potomac Associates, Earth Island Limited, London, 1972).

MITROFF, I., V. BARABBA, R. KILMANN, "The application of Behavioral and Philosophical Technologies to Strategic Planning: A Case Study of a Large Federal Agency". *Management Science*, Vol. 24, No. 1 September 1977.

OECD, Interfutures (Paris, 1979).

PREMFORS, R., "Social Research and Public Policy Making: An Overview", *Statsvetenskaplig Tidskrift*, 1979 (4).

Programme for Future Studies in Sweden (Secretariat for Future Studies, Stockholm, 1975).

QUADE, E., Analysis for Public Decisions (American Elsevier, New York, 1975).

RAVETZ, J.R., Scientific Knowledge and its Social Problems (Hammondsworth, Penguin Books, 1971).

ROSSINI, F., A. PORTER, "Frameworks for Integrating Interdisciplinary Research", *Research Policy* 8 (1979) pp. 70-79.

RULE, J.B., Insight and Social Betterment (Oxford University Press, New York, 1978).

SVEDIN, U., "Technology, Development, and Environmental Impact - - An Introduction to the Scenarios", *Ambio*, Vol. VIII, No. 2-3, 1979.

To Choose a Future (Ministry for Foreign Affairs and Secretariat for Futures Studies, Stockholm, 1974).

WHISTON, T., (ed), The Uses and Abuses of Forecasting (Macmillan, London; Holmes and Meier, New York, 1979).

WILSON, I., "Scenarios", pp. 225-247 in J. Fowles (ed), Handbook of Futures Research (Greenwood Press, Westford, Connecticut, 1978).

# 6 The Legitimacy of Futures Studies

One of the functions of policy-orientated futures studies is to provide linkages between knowledge and policy-making. In the case studies presented in Chapter 4 and in the previous sections of this chapter we have reviewed several experiences of such linkages. However, regardless of whether or not a linkage is successfully achieved, problems connected with the legitimacy of futures studies loom large. These problems are mainly of three types.

First, there is the problem of the intellectual legitimacy of futures studies: how can statements emanating from activities of this kind be justified in intellectual terms, and how can they be distinguished from guess-work and purely partisan argumentation? A major part of the present report has focussed on this problem. Generally speaking the thrust of the argument has been that futures studies can and should be conducted in accordance with at least some broad intellectual guidelines. During the last few decades there has been some development of methods and criteria. It has to be admitted, however, that the relative importance of these methods and criteria can vary between different types of futures studies, and even within a class of futures studies, such as the policy-orientated kind that we have focussed on here.

That people differ in their ideas about the nature and purpose of futures studies may depend to a considerable extent on the great diversity of the field and the variety of organizational solutions employed. At present some futures studies are conducted at universities, but more are undertaken by institutes engaged in applied research or by "think tanks", by consultancy groups or by groups in various organizations. It is natural that those who would like to see more activity in this field at the universities should stress the compatibility between the criteria of adequacy and quality for futures studies and the paradigms of the social sciences. Goldmann argues along these lines and makes a comparison based on three categories of criteria:

- value independence
- validity
- level of ambition (generality, cumulativeness, originality and creativeness). (Goldmann, 1978)

Goldmann discusses how these criteria can be interpreted for the field of futures studies, but he also points out that certain modifications are called for. Goldmann's views seem to agree fairly well with those expressed by Dror in his discussion of the paradigms of policy sciences (Dror, 1971, Chapter 8). Dror, who regards policy-orientated futures studies as a field within the policy sciences, argues that "despite all innovativeness, policy sciences belongs to the scientific endeavour and must meet the basic tests of science in respect to verification and validation". Policy sciences should not be judged by the paradigms of contemporary sciences, but it must meet the basic criteria of all scientific efforts, and, in addition, the not less demanding tests of saliency to the improvement of policy-making."

However, futures studies are often undertaken outside the universities. The focus on the future and the claim to exert influence on policy also gives this field of study a special character, and makes it natural that many people have emphasized the need for special paradigms. Amara (1978, 1980) is a representative of this view. In the preceding sections we too have discussed various requirements and criteria. But we cannot necessarily assume that these special paradigms will be easy to reconcile with the paradigms of traditional research. Thus, Dror (1980) has highlighted a potential dilemma faced by many policy-orientated futurists. They need considerable freedom in producing innovative studies, but they often depend heavily on clients for purposes of financing and access to information. And such dependence may well entail limitations on the scope and conduct of futures studies.

When the stimulation of the public debate is regarded as one of the functions of a futures study, the application of traditional research criteria leads to an obvious dilemma. In such a case the presentation must be easy to read, but it is difficult to combine this with a detailed discussion of assumptions, limitations, choice of perspective, etc. However, the dilemma is not necessarily unsolvable, since results can be presented in different publications. The publication policy of the Secretariat for Futures Studies, with its series of reports from each project group, can be seen as a way of trying to handle this problem. Another way of dealing with it can be to present a futures study in two parts or versions - one for the general reader and one including more methodology for specialists. This last approach was chosen, for instance, in a recent doctoral thesis at the University of Stockholm (Vavrin, 1980). In the Transportation Study methodological discussions are intertwined in the general text, which has probably made the report less easy for the general public to read than the Energy and Society and The Resources and Raw Materials studies.

In the three projects described in Chapter 4 a good deal of attention was devoted to surveys of the state of knowledge in a number of fields related to the policy problems of resources and raw materials, energy, and transportation. The three project groups also took care to spell out the assumptions underlying the alternative future developments that were described. The main criticism that can be levelled at these studies does not primarily concern what was actually accomplished; rather, it could be argued that at least in the Resources and Energy studies some alternative approaches were tacitly ruled out.

Thus, despite an attempt in the Resources Study to apply three broad theoretical perspectives, the reports of the project demonstrate how difficult it is to motivate explicitly a chosen balance between different perspectives.

In the Energy Study technical and organizational factors, rather than economic or political for instance, are the crucial ones in the analysis. But valid as this approach may be, little express effort is made in the report to justify it.

Secondly, policy-orientated futures studies are directly or inindirectly related to the processes of policy-making and planning. And more often than not they are commissioned by one or more agencies or ministerial units. Project groups carrying out studies of this kind are called upon to make a research-based contribution, often of a transdisciplinary nature, to the work of planning or policy-making (cf. Dror, 1980). But this raises the problem of the advisory legitimacy of futures studies: under what conditions can futurists be accepted as analysts and advisors, and when are they likely to be dismissed as intruders whose mandates and allegiances are ill-defined? The Swedish experiences presented in this report hardly provide a clearcut answer to these questions, but they do indicate that the problem of the advisory legitimacy of futures studies can be addressed in several quite different ways.

One possibility is to embed futures studies firmly in an organizational structure, in which futurists and policy planners engage in a formalized dialogue with clearly identified roles. Futurists are then part and parcel of an organizational framework, and the people who use the output of the studies are clearly identified and able to voice their demands with respect to the orientation and presentation of the studies. This is essentially the solution which has been chosen in Sweden's defence planning and which has come to serve as a model for planning systems in several other sectors of Swedish society. However, the Transportation Study presented here illustrates some interesting problems that can arise if application of this model is not preceded by the establishment of the appropriate organizational structure: in such a case implementation can cause problems, however policy-relevant the results of the studies may be.

The studies commissioned by the Secretariat for Futures Studies were intended to be relevant to a wide range of potential recipients: planners and policy-makers in government offices, political parties and interest groups, the academic community and, not least, the general public. In practice, the importance of different categories varied over time and between projects. Gradually the general public and certain non-establishment groups became more prominent than was originally intended as recipients of the results,to some extent at the expense of the government offices.

This development was largely a consequence of difficulties experienced by the Secretariat when it attempted to cut across traditional ministerial boundaries. The location of the Secretariat in a Prime Minister's Office whose coordinating role was comparatively weak, and the existence of a strong Ministry of Finance which did little to hide its suspicious attitude which sometimes bordered on overt hostility towards the very idea of a futures studies unit,aggravated this situation and made it even more difficult for the Secretariat to achieve advisory legitimacy in the ministries.

However, the case of the Resources and Raw Materials project group demonstrated that initial suspicion can be overcome, a certain legitimacy can be established, and good working relationships between futurists and several ministerial units can be maintained, without any loss in the ability to reach a variety of recipients including the general public.

But it is important not to equate advisory legitimacy with policy impact. The early impact of the results of the Energy Study was in fact a direct effect of the results being utilized by opponents of the then government's energy policy - or more precisely its nuclear policy - and of the corresponding loss of the project group's advisory legitimacy among many of the traditional energy policy administrators in the government office In a situation of redefinition and turbulence in the field of energy policy, an unorthodox posture on the part of the futures studies group enhanced rather than hampered its own visibility and impact on the general public debate and subsequently also on planning and policy-making. The challenging of established convictions is, after all, one of the chief missions of a policy-orientated futures study. But in carrying out this mission, a group may find itself moving from an advisory to an adversary position. And unless a unit enjoys strong top-level support and is more or less immune to attacks from the regular bureaucracy, a consistently adverse position might well threaten its own very existence (cf. Scharpf & Mayntz, 1975).

A third aspect of legitimacy concerns the problem of the societal legitimacy of futures studies.

Policy-orientated futures studies originated as part of a series of endeavours to handle the link-up between research and knowledge on the one hand and policy-making on the other. Numerous tags such as "PPBS", "social indicators", "sectoral research" and so on bear witness to the same pervasive "scientification" of policy-making and administration.

These endeavours were often accompanied by exaggerated hopes of achieving a more rational management of administrative and policy problems. But they also provoked opposition. An expanding "hybrid" sector inhabited by "technocratic" administrators and planners has been envisaged as a threat both to academic freedom and to the traditional virtues of bureaucracies. The politicization of science and the scientification of politics and administration has been regarded as a mixed blessing in both radical and conservative circles. Nor should such objections be lightly dismissed. But they show that the problem of the societal legitimacy of futures studies is not a problem of futures studies per se, but of futures studies in so far as they form part of various more encompassing developments in society.

One such development concerns science policy in Sweden and other Western nations. Policy-makers are trying increasingly to relate research to social and political objectives, possibly - so the critics argue - at the expense of a longer-term and freer-ranging search for knowledge unbound by considerations of direct utility and immediate returns. (For overviews of recent Swedish science policy, see Elzinga, 1980, and Wittrock, 1980).

Another feature of relevance to the problem of the societal legitimacy of futures studies concerns the explicitly political characteristics of various types of forecasting, futures studies and policy analysis. Questions such as these are closely linked to problems touching on the role of the state in contemporary society, and on what some observers see as the emergence of a new stratum of intellectuals and functionaries which is assuming a leading position and acquiring new élitist privileges (cf. Bruce-Briggs, 1979, Gouldner, 1979).

Obviously futurists cannot be expected to solve broad social and analytical problems of this kind. But a minimum requirement, which has been formulated by Jonathan Gershuny, can perhaps be suggested: futurists and forecasters should not let themselves be mistaken for neutral instruments in the politicians' hands (Gershuny, 1981).

REFERENCES

AMARA, R., "Probing the Future" In J. Fowles (ed) Handbook of Futures Research (Greenwood Press, Westport, Connecticut, 1978).

AMARA, R., The Futures Field (Institute for the Future, Menlo Park, California, Sept 1980).

BRUCE-BRIGGS, B. (ed), The New Class? (McGraw-Hill, New York, 1981).

DROR, Y., Design for Policy Studies (Elsevier, New York, 1971).

DROR, Y., "Think Tanks: A New Invention in Government", pp. 139-152 in C.H. Weiss and A.H. Barton (eds), Making Bureaucracies Work (Sage Publications, London and Beverly Hills, 1980).

ELZINGA, A., (1980), "Science Policy in Sweden: Sectorization and Adjustment to Crisis" *Research Policy*, Vol. 9, 1980, pp. 116-146.

GERSHUNY, J.L., "What should Forecasters Do? A Pessimistic View" pp. 193-207 in P.R. Baehr and B. Wittrock (eds), Policy Analysis and Policy Innovation (Sage Publications, London and Beverly Hills, 1981).

GOLDMANN, K., "Om kvalitetskontroll av framtidsstudier", *Internationella studier*, 1978:6.

GOULDNER, A., The Future of Intellectuals and the Rise of the New Class (The Seabury Press, New York, 1979).

SCHARPH, F.W., & R. MAYNTZ, Policy-Making in the German Federal Bureaucracy (Elsevier, Amsterdam, 1975).

VAVRIN, J., The Airline Insurance Industry - A Future Study (Ph D thesis, Department of Business Administration, Stockholm University, 1980).

WITTROCK, B., "Science Policy and the Challenge to the Welfare State", *West European Politics*, Vol. 3 (1980) pp. 358-372.

# About the Authors

BRITA SCHWARZ, Ph.D. Dept. of Mathematics, Stockholm University 1958, Director of Research in Systems Analysis and Planning at the Research Institute for National Defence. From 1967 to 1970 she was consultant to OECD (Paris) on questions of programme budgeting and long-range planning. Her publications in English include Budgeting, Programme Analysis and Cost-Effectiveness in Educational Planning (OECD, 1968), and contributions to Knowledge and Concepts in Futures Studies (ed. S. Schwarz, 1976), Perspectives on Research, Development and Technological Innovation (ed. D. Sahal, 1979), Applied Cybernetics and Planning (ed. A. Ghosal, 1980) and the IIASA Handbook of Systems Analysis (forthcoming). She participated in a futures research project in transportation between 1976 and 1978 and since 1980 she has been President of the Swedish Society for Futures Studies. She is currently associated with the Economic Research Institute at the Stockholm School of Economics (address EFI, Box 6501, S-113 83 STOCKHOLM, Sweden).

UNO SVEDIN, Ph.D. University of Stockholm 1974, at present Assistant Professor of Physics (specializing in Systems Analysis). In 1975 he held a Sweden-America Fellowship; his particular interest was futures studies and technology assessment. In 1975-1977 he participated in the project group "Resources and Raw Materials" at the Swedish Secretariat for Futures Studies and was co-author of Resources, Society and the Future, Liber, 1977 (in Swedish), Pergamon 1980 (in English). He was in charge of the preparation of the Swedish National report, and was a member of the Swedish delegation to the UN Conference on Science and Technology for Development (UNCSTD) in Vienna in 1979. In this connection he acted as guest editor of a special issue of AMBIO in which type scenarios for future developments in third world countries were presented. He is at present coordinator of a three-year project on "Natural resources in a cultural perspective" at the Committee for Future Oriented Research at the Swedish Council for Planning and Coordination of Research (Address FRN, Box 6710, S-113 85 STOCKHOLM, Sweden).

BJÖRN WITTROCK received his Ph.D. in 1974 from the University of Stockholm where he is now Associate Professor of Political Science and Chairman of the Group for the Study of Higher Education and Research Policy. He has held visiting positions at Washington University (1976) and the University of California at Berkeley (1979 and 1980). His recent publications in English include *Policy Analysis and Policy Innovation* (co-editor; 1981) and contributions to *The Uses and Abuses of Forecasting* (1979), *Internationale Dimensionen in der Wissenschaft* (1979), *Politics as Rational Action* (1980) and to such journals as *European Journal of Political Research*, *Futures*, *West European Politics* and *Journal for General Philosophy of Science*.